Friction

Friction

A BIOGRAPHY

Jennifer R. Vail

The Belknap Press of Harvard University Press

CAMBRIDGE, MASSACHUSETTS · LONDON, ENGLAND · 2026

EU GPSR Authorised Representative
LOGOS EUROPE, 9 rue Nicolas Poussin,
17000, LA ROCHELLE, France
E-mail: Contact@logoseurope.eu

Many of the designations used by manufacturers and sellers to
distinguish their products are claimed as trademarks. Where
those designations appear in this book and Harvard University
Press was aware of a trademark claim, then the designations
have been printed in initial capital letters.

LIBRARY OF CONGRESS CATALOGING-IN-PUBLICATION DATA
Names: Vail, Jennifer R. author
Title: Friction : a biography / Jennifer R. Vail.
Description: Cambridge, Massachusetts; London, England : The Belknap Press of
    Harvard University Press, 2026. | Includes bibliographical references
    and index.
Identifiers: LCCN 2025022619 (print) | LCCN 2025022620 (ebook) |
    ISBN 9780674290662 cloth | ISBN 9780674303867 epub | ISBN 9780674303874 pdf
Subjects: LCSH: Friction | Tribology | Technological innovations—History
Classification: LCC TJ1075 .V254 2026 (print) | LCC TJ1075 (ebook) | DDC
    621.8/9—dc23/eng/20250807
LC record available at https://lccn.loc.gov/2025022619
LC ebook record available at https://lccn.loc.gov/2025022620

For all the tribologists out there:

keep rubbing things the right way.

Contents

Friction

Introduction

THE PROBLEM WITH BEING a tribologist is that you're inundated with puns. Tribology, the science of interacting surfaces in relative motion, is concerned with the friction, wear, and lubrication of materials. As an experimental tribologist, it is my job to rub things together. You wouldn't want to *rub* them the wrong way because that's a slippery slope. Perhaps, you see how this could start to *wear* on you?

The other problem with being a tribologist is that you can't escape it. Tribology is all around us. Even when I am ready to call it a day at work, those darn interacting surfaces persist all around me. Just driving home, I experience a wealth of tribological interactions: my foot on the pedals, the tires gripping the road, not to mention all the moving components in the engine.

The truth is, neither of these problems is a problem at all. They highlight the beauty of tribology: once we discover it, we realize that it's fundamental to who we are. Humans have always been tribologists. Our ability to manipulate friction is at the heart of civilization, from the discovery of fire in the Paleolithic era to the invention of nonstick pans in the twentieth century. You, too, have adapted to and tinkered with friction your whole life.

This brings us to one of the most common questions I'm asked: How did you become a tribologist? I hadn't even heard the word *tribology* until university. When I told friends and family I was pursuing a PhD in it, the reactions included wondering how the little *Star Trek* critters amounted to an entire PhD, and speculating why I had studied mechanical engineering only to pivot into anthropology. My path to tribology was simple: I like mechanics, I like materials, and tribology lives at the intersection of both.

Once I began studying tribology, I realized that destiny had played a role. I began my mastery of manipulating friction during my years of dancing. En pointe, I used rosin to prevent my shoes from slipping on the floor, quickly learning just how much to use for optimum performance. My tap shoes were equally concerning since no dancer wants to wipe out on stage. I scuffed the bottoms of the shoes near the taps, achieving just enough friction to gain traction and still glide effortlessly across the floor.

I've since hung up my dancing shoes, but I still use rosin to create friction between my bow and the strings of my viola. And yes, colleagues have studied the friction of sound because that's the beauty of tribology: it's everywhere. After all, as tribologists, what can't we study? In the span of a few years, I worked on aerospace composites, then syringes, then pet food. Why? Well, we need aerospace materials that are lightweight, long lasting, and efficient. The plunger of a syringe must pass smoothly through the barrel or getting vaccinated would truly be an ordeal. Pet food may sound like a detour, but our pets develop plaque just like us. What if we could develop food and treats that removed plaque just by rubbing against our pets' teeth? That is how my dentist dad and his tribologist daughter finally spoke the same language. I even read the latest versions of the textbooks he used in dental school.

Each of these projects touches our daily lives in some way. Tribologists around the world work in seemingly niche subjects that collectively influence the world around us. From the invention of fire and artificial implants to improving fuel mileage on cars and designing so-called frictionless hyperloops, each tribological discovery has pushed us to reimagine what technology can be. The way we experience the world, whether through greater efficiency, flight, or space exploration, has been shaped by our understanding of friction. Probing the laws of friction has not only given us insight into myriad biological and physical mysteries, it holds the promise of helping us solve one of the most urgent challenges of our times—the energy and climate crisis. And, when we push beyond Earth, friction may even help us understand the evolution of planetary systems and the origins of the universe.

1 *Tribology Is Born*

IN 1964, spirited debates erupted at a conference in Cardiff, Wales. The source of the consternation was extreme production delays that stemmed from malfunctioning equipment at iron and steel plants across the United Kingdom. Beginning in the twentieth century, the automation of heavy machinery enabled plants to operate continuously, increasing productivity and revenue. The downside was that any small hiccup was acutely felt, cascading through the production line.[1] At first, it was assumed that inadequate lubrication of factory equipment was causing parts to seize up or break apart. And so, the Lubrication and Wear Group of the Institution of Mechanical Engineers, along with the Iron and Steel Institute, called on engineers and representatives from industry to convene and get to the bottom of the problem. But as participants examined photos of equipment failures at plants across the United Kingdom, United States, and Germany, they realized that the problem was not in the lubrication but in the design of the equipment itself. The old designs weren't equipped to handle the wear and tear of continuous operation, no matter how much lubrication was applied.

In attendance at the conference was Peter Jost, an engineer from the British iron, steel, and tinplate production plant Richard Thomas and Baldwins. Conveniently, one of Jost's acquaintances, Bertram Vivian Bowden, Baron Bowden, had recently become the UK's Minister of State for Education and Science. Jost relayed the findings to Lord Bowden and suggested that the newly established Ministry of Technology look into the matter further. When Lord Bowden's inquiries to the Ministry of Technology went unanswered, he decided to take matters into his own hands. He assembled a team of engineers to investigate the scope of the failures and to make recommendations for addressing them. Jost would lead the team.

As the engineers collected data from dozens of plants in the UK, Europe, and the US, a pattern emerged. During operation, as surfaces rubbed against each other, an inadequately designed system experienced excessive wear and tear as well as energy losses. In the UK, these inefficiencies amounted to over £500 million in lost revenue, approximately 1.3–1.6 percent of the gross domestic product (GDP).[2] Designing a properly lubricated system that could overcome such inefficiencies, they argued, required an interdisciplinary approach informed by mechanical engineering, physics, and chemistry.

By late 1965, Jost's team was ready to present its findings to the Department of Education and Science. There was just one problem. The title of the report, "Lubrication," didn't convey the interdisciplinary nature of the investigation they had undertaken. The marriage of chemistry, physics, and engineering to study how surfaces interact was novel and significant enough to be considered a new discipline of science, and it needed a name. As Jost noted, the team's description of this discipline, "Interacting surfaces in relative motion and associated practices," was a mouthful.[3] Hence, some rebranding was in order.

Jost reached out to the editor of the *Oxford English Dictionary*, Robert Burchfield, who pointed out that rubbing was the commonality among all the failures. He suggested using the Greek word *tribo,* meaning "to rub," as the root of the name. After all, one of Jost's consultants, David Tabor, had already co-founded the Laboratory of Tribophysics in Melbourne in the late 1940s. And the Dutch scientist Petrus van Musschenbroek had used the term *tribometer* to describe an instrument he invented to measure resistance in bearings.[4] The editor proposed "triboscience and tribotechnology," or, if Jost preferred something shorter, "tribology." At first, not everyone was on board. But eventually they agreed that a word rooted in ancient Greek would be more easily adopted by an international and interdisciplinary field. Many scientific terms have Greek roots, thanks to the prolific discoveries of the ancient Greeks.[5]

In November 1965, the report, titled "Lubrication (Tribology)," was sent to the Minister of State for Education and Science, a position now held by Edward Redhead. The response was swift and comprehensive.

Within the Ministry of Technology, a Committee of Tribology was formed, once again headed by Jost. On this committee, government employees joined academics and industry experts to advise the government on how to implement the findings of the report, including how to provide training in tribology. Soon, centers for the study of tribology emerged across the UK, establishing deep expertise in the science of surfaces and placing it at the forefront of engineering research. Today, Jost is widely known as the father of modern tribology. We have him to thank for labs and doctoral degrees devoted to wear, lubrication, and the force that opposes motion.

Or, as we know it, friction.

Friction in Ancient History

Mirroring its linguistic origins, tribology has its roots in ancient history. Our relationship with friction dates back to one of humanity's greatest discoveries: fire. Exactly when humans mastered fire has long been debated. Conclusive archaeological evidence, such as burned bones or ashes, can be difficult to find. Until recently, it was believed that our ancestors, *Homo erectus* and *Homo neanderthalensis,* learned to control fire around 400,000 years ago. Then in 2013, researchers at the Wonderwerk Cave in South Africa, one of the oldest known human settlements, uncovered million-year-old ash thirty meters from the cave's entrance, well away from where any natural fire would have started. It appeared that the hominids who occupied the cave a million years ago had managed to bring fire into their home. Whether the residents of Wonderwerk Cave mastered fire to the extent that they could create it on demand or whether they merely understood how to maintain a long-lasting, steady flame, humans may have had fire much longer than originally thought.[6]

But what does friction have to do with fire? Fire requires three ingredients: fuel, oxygen, and heat. Fuel can range from wood to straw to any combustible material, such as gasoline. The combination of these materials affects the chemistry of the reaction. Oxygen is readily available in the atmosphere. Heat? That's where friction comes in. The act of rubbing objects together creates a form of energy called *mechanical energy.* The motion produces friction, which converts some of the mechanical energy into *thermal energy,* also known as heat. Once there's enough thermal energy, the atoms in the fuel combust, combining with oxygen to form a vaporous gas—fire. Since different materials have different heats of combustion, the duration, speed, and effort used in the rubbing affect how readily a fire starts. The more friction our ancestors could create by rubbing a stick against wood or a stone, the faster the fire would have started.

Charles Darwin considered fire one of humanity's greatest discoveries.[7] Our mastery of fire enabled permanent settlements and ignited (pardon the pun) a remarkable number of technological innovations, from cooking to pottery to metalworking. Fire also shaped our evolution. Cooked food, easier to chew and digest, allowed for more efficient extraction of nutrients, and as a result, our digestive tracts shrank, and our brains grew larger. All of this is thanks to our ability to control fire with the aid of friction.[8]

Beyond fire, the ability to manipulate friction has fueled many of humanity's most significant cultural achievements, including such ancient marvels as the Egyptian pyramids. The Pyramid of Giza, begun around 2550 BC, boasts an estimated 2.3 million stones, weighing six tons in total. At its base, each stone measures one meter in length, two and a half in width, and one and a half meters in height—a staggering size for materials moved with human labor.[9] How did the Egyptians move these mammoth stones centuries before the advent of the wheel? The question flummoxed researchers for years, until a team of physicists at the University of Amsterdam realized in 2014 that the Egyptians had already shown them the answer in a wall painting on the tomb of Djehutihotep.[10] During the Twelfth Dynasty, Djehutihotep presided over the district that included Hermopolis, at the border of Upper and Lower Egypt. At the time of Djehutihotep's reign, circa 1900 BC, Hermopolis was a thriving metropolis, its opulence surpassed only by Thebes.[11]

The painting depicts 172 men hauling a large statue on a sledge across the desert. The colossus, yet to be uncovered in excavations, is believed to have stood nearly seven meters tall and weighed at least sixty tons.[12] Even with so many men to divide the weight equally, they'd each have to haul 350 kg or 772 lbs. To put this in perspective, a baby elephant weighs about one hundred kilograms. The last thing these men needed was friction working against them.

The clue to how the Egyptians might have moved these heavy blocks can be found at the front of the sledge, where a man is seen tipping a basin of water toward the ground. For years, historians believed the scene to depict a ceremonial act. But if so, it was a ceremonial act that would have eased the pain of moving heavy objects across the desert. To test this idea, the physicists decided to recreate the scene by building miniature sledges, which they pulled across trays of sand. If you've ever pulled something through dry sand, you know that sand is loose and piles up around the object you're trying to move, increasing the resistance. When the researchers added water to the sand, they found what the ancient Egyptians must have realized, even if accidentally, millennia ago: adding small amounts of water made the sand stiffer, preventing heaps of sand from forming and reducing friction. But as the Amsterdam team discovered, and likely the Egyptians before them, it's not as simple as just adding water. The *amount* of water matters. Water bridges that form between sand particles will begin to merge and disappear as the mixture becomes saturated, reducing the mechanical enhancement they provide.

The Egyptians were not the only civilization to grapple with more efficient transportation. The Romans are perhaps the most famous example, with their renowned chariots. They employed a tribological trick that we still rely on today: rolling to reduce friction. We often forget that the first wheel was used not for transportation but to make pottery. It enabled more efficient and uniform production of ceramic ware, which was essential for cooking and trade.[13] It wouldn't be used in transportation for another three hundred years, when it was incorporated into Roman chariots. The fact that so many machines still rely on the wheel is a testament to this simple invention's astounding success. Friction was key to that success, and in more ways than one.

A wheel needs friction to move. This might come as a surprise, given that we just saw how friction resists motion. As Isaac Newton explained in the seventeenth century, every action has an equal and opposite reaction. At rest, the weight of a wheel exerts a force against the ground. The ground, in turn, pushes back. This resulting force is a component of friction called *static* friction, because at this point, movement hasn't commenced. When we exert a force to move the wheel forward, the force between the ground and the wheel keeps the wheel from spinning freely. It enables the wheel to push off and move linearly as opposed to continuing to rotate in place. As contradictory as it sounds, friction, the force that resists motion, also enables the motion of the wheel by pushing it forward.

Once forward motion begins, the resistance to motion becomes *dynamic* friction, which is just as important to the operation of a wheel. If you've driven on ice or experienced hydroplaning in a car, you know the danger of too little friction. Of course, if there's too much friction, static or dynamic, between the wheel and ground, the wheel will get stuck and go nowhere. Wheels must be designed to achieve just the right amount of friction.

Early wheels were rather primitive, made of simple, uneven wooden disks sliced from tree trunks. This can give the false impression that ancient civilizations were accidental tribologists rather than intentional ones. However, the Romans' solution to the other critical friction issue with a wheel—that it must rotate *around* something—tells us otherwise. To function properly, the wheel must rotate around an axle, or shaft. This is why the wheel is often referred to as a unit: the wheel and axle. And where there is motion, there is friction available to oppose it. If friction between these components is too high, the wheel can't rotate at all, or will struggle mightily to do so.

Prior to the chariot, potters solved the problem of friction created by the motion of a wheel around an axle by working with wet clay. Clay needs to be wet to be malleable, and water is an excellent lubricant that makes things slippery. At first, the Romans tried lubricating the axle bearings with water. The problem with water is that it doesn't hang around very long, making it an impractical solution without an injection system to keep the wheels lubricated.

Fortunately, the Romans were not the first civilization to grapple with this problem. The Sumerians had already realized that fats and oils could be used as lubricants, a practice that the Romans wasted no time in adopting. Circa 160 BC, Cato the Elder wrote in *De Agri Cultura,* a treatise on agriculture, that amurca, a byproduct of olive oil production, was useful for lubricating leather and saddles.[14] Nearly two hundred years later, Pliny the Elder mentioned the practice of lubricating the axle of a wheel in *Naturalis Historia.* The largest surviving text from the Roman Empire, this thirty-seven-volume work covers scientific topics ranging from astronomy to botany. Pliny noted that "the ancients" would apply lard to wheels to make them "revolve more easily." When the lard was mixed with the rust from wheel hubs, Pliny explained, the material developed medicinal properties. He called the product *axungia,* meaning axle grease, because of the lubricating properties of the applied lard.[15]

Greasing the axles did more than enable movement of the wheel. It helped reduce wear and damage and provided an important bonus for chariot racers: reducing the squeaking of parts. We've all used WD-40 on our squeaky door hinges. Axungia was the ancient equivalent.[16] There has been much debate around the relationship between friction and noise, but reducing friction generally reduces noise. Perhaps the Romans didn't fully appreciate that their lubrication practices resulted in quieter chariot operation by reducing friction, but they certainly understood that easier motion was quieter.

Noise wasn't the only discomfort friction caused. We know that rubbing causes surfaces to heat up, a fact that we have used to our advantage for millions of years—since we first created fire. But as advantageous as this can be when we have cold hands, it's not always desirable. Chariots used in racing, with their fast-moving parts, endured a fair amount of frictional heating. In *Odes,* a collection of lyric poems, the Roman poet Horace described how chariot wheels would glow during racing.[17] Chariots raced multiple times a day, at speeds up to forty miles per hour (mph). Frictional heating was enough to char the wheel hubs, and that heat had to go somewhere. Unfortunately for the driver, that heat would be absorbed by the floorboards of the chariot, overheating the driver's feet and warping the wood in the process.

Much like Formula 1 (F1) racing today helps push the bounds of automotive engineering, ancient chariot racing did the same for tribology. The challenge came from what should have been the solution to friction: lubricating the wheels with grease. Greases have an ignition point, which means they will catch fire if their temperature exceeds that point. High-speed races could generate enough frictional heating to ignite grease. The Romans needed a cooling system. For this, they reverted to water, rehydrating the wheels after a certain number of laps. Evidence of this practice can be seen in mosaics, which depict amphoras of water either being poured over the wheels or placed around the track.[18]

And thus friction gave rise to the pit stop as we know it.

If we look to the seas as a means of travel, the Vikings reigned supreme, voyaging as far east as Greenland and North America. The Vikings had a sophisticated understanding of how to vary design for function, as exemplified by the diversity of their ships. You know you have a robust fleet when you can sink a few to ward off an enemy, as the Vikings did with the Skuldelev ships. Discovered in 1962, these five ships, dating back to around AD 1030, range from cargo carriers to warships. They were likely used to block a channel west of Copenhagen from invaders.

When speed was of the essence, the Vikings built narrower hulls that reduced the surface area in contact with water, minimizing friction, also known as drag. At thirty meters long and just 3.8 meters wide, Skuldelev-2, a warship and one of the iconic longships, could reach speeds up to seventeen knots, or nearly twenty mph. However, when the Vikings needed to transport heavy goods between ports, they chose a wider hull that could provide stability and strength at the cost of increased drag and reduced speed.[19] Skuldelev-1, a cargo ship, was sixteen meters long, five meters wide, and maxed out at thirteen knots, or fifteen mph. An understanding of the relationship between stability and drag doesn't appear to have been accidental. Archaeo-logical evidence shows that Norwegian Viking longships, more likely to encounter rougher seas, generally had wider hulls than Danish ships. Transportation has a way of making us contend with friction through clever engineering solutions.

Da Vinci, a Founder of Tribology

If Jost is considered a founder of tribology, then he is joined at this elite table by none other than Leonardo da Vinci. Between 1482 and 1499, da Vinci was employed by the Duke of Milan as a military engi-neer, architect, sculptor, and painter. Da Vinci was fascinated by the idea of perpetual motion. The lure of a wheel that could spin indefinitely without an external energy source dates to the twelfth century in India, when engineers first tried their hand at designing such a ma-chine.[20] Da Vinci devoted pages of his notebooks to the idea, before realizing that his efforts were futile. Friction, he determined, made perpetual motion impossible.[21]

Friction was not a new concept in da Vinci's time. As we've seen, civilizations had been manipulating and learning to coexist with it for millennia. About two thousand years before da Vinci, Aristotle con-sidered resistance to motion a property of the medium through which objects moved. In *Physics,* a compendium of lessons on the philosophy of causality and principles of the natural world, Aristotle presented his ideas about forces and motion. Today, we refer to these ideas as Aristotelian physics. The velocity of an object, according to Aristotle, was related to the force applied to that object and the resistance of the medium through which it moved. Aristotle was on the right track, although he also claimed an object must experience an applied con-stant force to achieve a constant speed. He believed that the force needed to sustain the motion of an object came from air rushing around it. That meant two balls of different weights would fall at dif-ferent speeds since the heavier ball would have more air pushing on it.[22] If you've let a ball roll down a ramp, or dropped two objects from high enough, you know this isn't true.[23] These early ideas on resistance to motion led to little advancement in the understanding of friction. It was not until da Vinci came along that the first known attempts to quantify friction were made.

Da Vinci's notebooks are filled with detailed sketches of complex contraptions, from parachutes to helicopters. He knew from observing the world around him and tinkering in his workshop that forces act against motion. This was a limiting factor in designing machinery, and da Vinci became fixated on tackling it. But first, he needed to understand how this force worked. Over the next decade, da Vinci worked tirelessly on this problem, leaving behind the tribologist's equivalent of the *Mona Lisa:* a sketch of the first tribometer, an instrument that could measure friction. Tribologists use tribometers for a variety of problems, from measuring the friction of the soles of our shoes to developing a lubricant for a gear system. Da Vinci's device consisted of three blocks pulled by a cord connected to a pulley system. Attached to the end of the cord was a circular weight used to apply force to the cord and pull the blocks. With this primitive tribometer, da Vinci could quantify the friction in the system based on the amount of weight required to move the blocks across the surface. Stacking blocks vertically and horizontally, he could determine how their weight and orientation influenced friction. This led to a key finding: friction is affected by the load, or weight, of the blocks and not, as one might expect, by the apparent amount of contact area between the blocks and the surface. Scribbling in red ink on a scrap of paper, he noted that friction "is of double the effort for double the weight."[24]

The proportionality between friction and the load applied perpendicular to the direction of motion is called the *coefficient of friction*. It's arguably one of the most important concepts in tribology.[25] From the value of the coefficient of friction, we can deduce how much resistance there is in a system. The lower the coefficient of friction, the less force you need to move that system, but the higher it is, the more resistance you encounter. Over the next few years, da Vinci revised this ratio numerous times, ultimately arriving at a value of .25.

Today, we know that there is not just one value for the coefficient of friction. And at some point, as da Vinci repeated these experiments

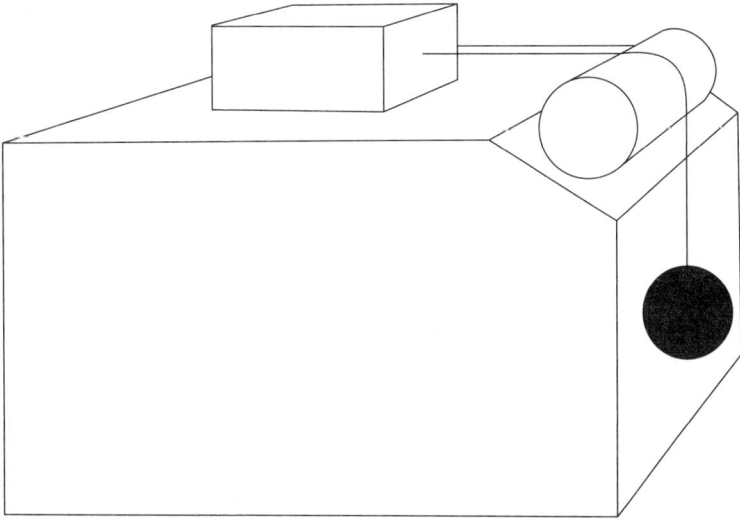

Schematic of Leonardo da Vinci's early tribometer

with materials ranging from soaps to lubricated surfaces, he realized this as well. In fact, he ultimately referred to the coefficient of friction with what would become a favorite statement of tribologists: "It depends." Da Vinci understood that different types of materials create different amounts of friction. Hardness and surface roughness, he noted, made objects easier or more difficult to move. When designing mechanical systems, material selection, including the surface finish of the material, mattered. Da Vinci, in the early sixteenth century, was beginning to explore a field developed centuries later, one that we'll explore in Chapter 2: contact mechanics.

For nearly a century, da Vinci's insights remained buried in his notebooks. Da Vinci didn't publish these findings, though he did document them. It may be that he never felt his work was complete and ready for publication. His notebooks have revealed that he took a meticulous approach to experimentation: observing the world around him, developing hypotheses, testing through experimentation, and documenting his findings. The latter step has proven to be quite a gift to historians, if a somewhat frustrating one. Da Vinci developed a shorthand and wrote in mirror image, perhaps because he was left-handed and wanted to avoid smudging the ink—something to which lefties everywhere can relate.[26] It would be another century until Amontons, a French inventor and physicist, rediscovered what da Vinci had already found.

Amontons and the Laws of Friction

Born in 1663, Guillaume Amontons became deaf in adolescence. This limited his career prospects. Fortunately for physics, Amontons chose to devote himself to his studies, although his family initially opposed his desire to study physical sciences. He never attended university but was driven to study drawing, surveying, and architecture. Ultimately, his interests prevailed over his family's wishes, and he gained experience in applied physics as a mechanic for public works projects.[27]

Of all his areas of study, physics was his true passion. Amontons, like da Vinci, was captivated by the prospect of building a machine that could run on its own. Unaware that da Vinci had shown perpetual motion to be impossible because of friction, Amontons devoted considerable attention to perpetual motion machines. His efforts led to an invention he called the fire mill, a wheel that used the expansion of heated air to generate power and rotate. In 1699 Amontons presented the design to the Royal Academy of Sciences in a publication titled "A means of conveniently substituting the action of fire, for the strength of men and horses, to move machines."[28] The paper's title, somehow both concise and long, was a significant contribution to the field of *thermodynamics,* the study of the relationship between heat and other forms of energy. Amontons had converted heat to mechanical energy, developing what would become the steam engine, and eventually, the internal combustion engine.[29] It was through his work quantifying the mechanical energy needed for the mill that Amontons found himself exploring friction.

To estimate the amount of power his fire mill required to run without horses and men, he needed to quantify how much mechanical work people and horses could do. *Mechanical work* is the amount of energy transferred by a force and is measured by that force over a displacement. Amontons would first have to determine how much force a person would need to exert on the machine to move it a known distance. This force would be equal and occur in the opposite direction of friction.

Amontons began with polishing experiments, an exercise almost every novice tribologist conducts at some point in her training. Polishing involves moving a sample over a surface covered in an abrasive material to alter the surface finish of the sample. Amontons slid glass blocks over a glass plate covered in sand and emery. He chose polishing because he believed friction stemmed from surface roughness. To enable movement, one surface would either need to wear away or be deformed.[30]

But before he could determine how much mechanical work was necessary to move the glass blocks, Amontons had to solve a technical problem. He needed to isolate the horizontal and vertical forces in his setup. The mechanical work he was trying to quantify was based on the horizontal sliding distance of the sample. If you've ever tried polishing something, you know that you need to exert a downward vertical force to keep the object you're polishing in contact with the polishing medium. As the results of my own polishing exercises attest, it's not so easy to apply a vertical force while pushing something horizontally. Amontons's solution was to create a mechanical holder consisting of a spring that would press the block onto the glass plate. This enabled him to apply a small downward force to the block while pushing on its side with a horizontal force. Using a spring balance, he measured the horizontal force and the distance the block traveled on the table, enabling him to quantify the work done to move the block. By having people move the blocks repeatedly over time, he could calculate their rate of work, also known as power. Human output, he found, was about one-sixth the power of a horse. With this knowledge, Amontons could determine how his mill design stacked up against a horse or human-powered mill.[31] It would take the equivalent of 39 horses or 234 men to power the machine.[32]

By separating the horizontal and vertical forces, Amontons could measure the downward force that the block and spring exerted on the plate. This force, perpendicular to the direction of motion, is called the *normal force*. He then applied the normal force to three blocks of varying size, expecting that the blocks with larger surface area would have higher friction. Instead, the resulting friction was nearly identical. This meant that the frictional force, which was equivalent to the horizontal sliding force, was proportional to the normal load and independent of the apparent area in contact between the sliding surfaces. This was conceptually identical to what da Vinci had discovered more than two centuries earlier, the difference being that Amontons shared these findings with the world through letters, lectures, and publications.

Today these findings are known as the first two laws of friction, and they're stated as:

1. The force of friction is directly proportional to the applied load;
2. The force of friction is independent of the apparent contact area.[33]

Amontons's ideas were met with skepticism from the scientific community. The young, self-taught engineer had proposed overturning what was taken to be common sense—that friction increased with more contact between surfaces. Absurd! Today, however, these are among the most important laws in a tribologist's toolkit. In my own lab, these laws have been critical to developing new materials that can improve the fuel efficiency of aircraft engines, which, as we'll see in Chapter 5, is a crucial step in reducing greenhouse gas emissions.

Coulomb and the Third Law

The third and, to date, final widely accepted law of friction comes to us from a French engineer named Charles-Augustin de Coulomb. Born in 1736, Coulomb's work bridged mathematics and experimental approaches. He is renowned for his work on electromagnetism and is also widely considered the founder of contact mechanics. The most famous of Coulomb's laws isn't even related to his work on friction. Rather, it deals with the forces between charged particles. Because of this law, the SI unit for electric charge is, appropriately, called the Coulomb.[34] My professors used to say that you know you've made your mark when you have a theory or law named after you and that you've *really* made your mark when you get a unit named after you. Coulomb, clearly an overachiever, managed all of the above.

Coulomb was an engineer in the French military. Amusingly, none of his superiors in the School of Military Engineering had expected much from him. They noted that while he was intelligent, he wasn't likely to work his way up the ranks or have a notable military career.[35] That he even enrolled in the military was a bit of happenstance. His mother, descended from a wealthy and well-known family, insisted that her son study medicine. Much to her chagrin, Coulomb's passion lay in mathematics. Eventually, reality set in, and Coulomb, accepting that mathematics was unlikely to pay the bills, joined the military.

In 1777 the Academy of Sciences in Paris sponsored a prize for solving sliding and rolling friction problems and applying them to navy machines. The navy wanted to optimize friction in the design of mechanical systems such as pulleys and slipways, which were used to launch ships. Applying tallow was the preferred method for reducing friction, but it was less than ideal. Sometimes, the heat generated by friction would cause the tallow to act as an adhesive instead of a lubricant. The navy wanted more reliable solutions, and achieving such solutions started with understanding friction better. Two years later, no winner had been selected. The prize money was doubled, and Coulomb decided to tackle the subject of friction himself.

Coulomb worried that Amontons's experiments represented too small a subset of physical conditions to address real-world problems such as those the navy hoped to solve. Believing larger experimental setups were necessary to understand the behavior and influence of friction, he built different-sized blocks and loaded them with weights up to 2,500 pounds.[36] Whereas Amontons slid blocks a few inches at time, almost exclusively measuring static friction, Coulomb slid materials up to four feet. He repeated these experiments with various materials and designs: woods and metals, flat blocks and blocks mounted on rollers, and blocks coated in various animal fats. He ran hundreds of experiments. For the most part, these experiments confirmed Amontons's laws. Yet they also provided another crucial insight into friction. Most of the time, friction was independent of how quickly Amontons slid blocks across a surface. Today, this observation is known as the third law of friction and is expressed as:

3. Friction in a system is independent of velocity.

In 1781, Coulomb submitted his essay summarizing his findings on friction and machines, winning the Academy of Sciences award. He later expanded this essay into a book called *The Theory of Simple Machines* (*Théorie des machines simples*).[37]

Coulomb translated his work into the language of science: mathematics. He set Amontons's first law of friction to a formula, reminding us that friction is fun. Or rather, $F = \mu N$, where F is the force of friction, N is the applied normal load, and μ is the coefficient of friction. If you search for this equation online, you may see it ascribed to any of the three founders: da Vinci, Amontons, or Coulomb. But the name most commonly associated with this equation is that of the man who derived the equation, Coulomb, and so the friction force in the equation is referred to as Coulomb friction. Personally, I just call it the Fun Friction equation, but whatever you choose to call it, it's arguably the most famous equation in tribology. It's used in almost every engineering design that involves moving parts. Roller coasters, cars, rockets—they would fail to operate if engineers did not incorporate the third equation into their designs.

But some people say laws are made to be broken, and the laws of friction are no exception. That we now consider Coulomb's observation a law would likely have surprised him, given that even he noted exceptions to it. As we'll see in Chapter 4, a surefire way to violate it is to introduce a fluid such as water or air to the system of interest. The faster you drive, the more air resists the car moving through it. Another way to violate the law is frictional heating. Rubbing a surface fast enough can heat it up, causing it to melt or, as the Romans discovered when greasing their chariots, to ignite, drastically altering the friction in a system. In Chapter 3, we'll see how frictional heating arises, enabling us to ski and ice skate. Sometimes, frictional heating can alter friction by inducing chemical changes in a material, such as accelerating the formation of rust. When tribologists need to design for high speed—for instance, when working with motors or skis—we often use frictional heating to our advantage, choosing materials that will experience localized frictional heating and yield lower friction at those speeds. The third law of friction is better stated in this way:

3. Friction is independent of velocity for dry systems at relatively low speeds.

The first law holds for dry systems, for the most part. But while friction always depends on the applied load, it isn't necessarily proportional for all systems. The materials matter—whether they are hard, soft, or lubricated. And despite the second law, sometimes contact does influence the amount of friction. The reason for this has to do with the mechanics of contacting surfaces. Up until this point, our discussion of friction has focused on the macroscopic behavior of friction. If you've ever assembled Lego pieces, you know that the smaller they get, the trickier they are to handle. Friction is no different. At the microscopic scale, properties such as material hardness or roughness alter how surfaces interact with each other. This brings us to contact mechanics, a topic worthy of its own chapter.

2 *Friction*
Friend or Foe?

FRICTION, frankly, gets a bad rap. It is generally associated with difficulty and complication. It *resists,* making it seem counter to our goal of moving forward. Friction is considered a nonconservative force, meaning that energy is lost during motion. Of course, losing energy comes at a cost: lower gas mileage in cars, decreased efficiency and output of machinery, and the costs of trying to overcome energy losses.

Fortunately, tribologists push past this, or rather, lean into friction *because* it's a challenge. We can't just ignore friction. In many cases, we need friction. It may be a nonconservative force, but one of the guiding principles of physics is the law of conservation of energy. It reminds us that energy isn't really lost. Rather, mechanical energy, the energy needed to keep an object in motion, is converted to thermal energy, or heat. That same heat gives us fire (Chapter 1). We've seen how human evolution and culture have been deeply entwined with our understanding of friction. But in truth, most people don't think about friction until they must, like when their car skids on an icy road. Anyone who has been in that situation can appreciate that friction can, in fact, be beneficial. Our relationship with friction depends on what we are trying to accomplish. To use it to our advantage, or to work around its disadvantages, requires an understanding of how friction works and exerts its influence. That requires a closer look at mechanics and contacting surfaces.

Mechanics is the branch of physics concerned with forces and the motion of objects. Much of our understanding of the physical world stems from it. The mechanics of macroscopic objects is referred to as *classical mechanics,* while the mechanics of bodies, such as atoms and subatomic particles, is referred to as *quantum mechanics.* One of the foundations of classical mechanics is *inertia,* the property of matter whereby an object at rest will remain at rest or in motion until an outside force acts on it.

It was Galileo Galilei who introduced the concept of inertia. Galileo followed in da Vinci's footsteps, seeking to understand the world around him. Like Coulomb, his family pushed him to study medicine, the more lucrative career option for one inclined toward math and science. And just as Coulomb would over a century later, Galileo, too, convinced his father to allow him to study mathematics and natural philosophy.

The Church expected Galileo to follow Aristotelian physics and geocentric astronomy, which held that Earth was the center of the solar system and that the Sun and other planets revolved around it. Secretly, however, Galileo was a supporter of Copernicanism, the controversial heliocentric model of the solar system, having the sun at its center. Galileo had heard reports of a new invention, an early telescope that magnified objects up to three times their size.[1] After building his own version of this instrument and improving its magnification to twenty times an object's original size, Galileo began observing the night sky. He noticed that Venus, like the Moon, had phases. Its sunlit side would start as a crescent and gradually become a full disc before beginning the cycle again.

Galileo realized his observation of Venus provided proof for the heliocentric model. According to the geocentric model, Venus would always be the same distance from Earth. Since it was located between Earth and the Sun, we would observe only a small range of phases. Most of the illuminated side of Venus would face away from Earth, so that anyone observing from Earth would never see more than half of Venus, which isn't the case. This combination of phases and size meant that Venus and other celestial bodies, including Earth, must orbit the Sun.

But there remained one puzzling problem with heliocentrism: we don't feel Earth moving. According to the model, the planet not only moves around the sun but also spins on its own axis, creating a cycle of day and night. To explain this fact, Galileo introduced the concept of inertia. As we saw, Aristotle believed a constant force would be required to get an object in motion and have it stay in motion. Galileo revised Aristotle's view of motion, stating that a body in motion would remain so unless it was disturbed by an outside force. If you were to close your eyes in a car moving at a constant speed, you wouldn't realize you were in motion until the driver stepped on the brakes. Similarly, we don't feel Earth's movement because we're standing on Earth, moving with it at a constant speed as it rotates around its axis and revolves around the sun. If, however, an outside force were to suddenly give Earth a shove, we would become very aware of the motion of our planet.

To demonstrate the concept of inertia, Galileo rolled balls down inclined planes. According to his theory of inertia, he might have expected that the ball would keep on rolling indefinitely. But as we know, the ball eventually stops. Galileo realized that there was an external force acting on the ball. Depending on the materials he used, the ball stopped at different distances from the ramp. A force was acting to retard the ball's motion, and that force—friction—depended on the other materials interacting with the ball. Without friction, the ball would have rolled on forever.

Not everyone was a fan of Galileo's new ideas. In 1616 Galileo found himself at the center of the Roman Inquisition when the Copernican theory of heliocentricity was declared heretical. Galileo was forbidden from disseminating such views. The episode could have ended Galileo's investigation of inertia, but in 1623 Cardinal Maffeo was elected Pope Urban VIII. The new pope was a longtime friend and supporter of Galileo's. This change in circumstances gave Galileo, who was by then suffering from poor health and likely feeling the sense of urgency that can accompany it, the confidence to continue his heretical pursuits.

But in 1632 he once again found himself in trouble with the Church, this time for publishing a polemic against the Ptolemaic system upon which Aristotelian physics was based. In *Dialogue Concerning the Two Chief World Systems,* two philosophers and a layman debate geocentricity and heliocentricity. The Copernican philosopher, Salviati, presents Galileo's views. Sagredo is a neutral but sympathetic party. And then there's Simplicio, who argues the side of Aristotle. His name alone leaves no doubt about Galileo's views on Aristotelian physics. Even his supporter, Pope Urban VIII, couldn't ignore the brash move by Galileo. Condemned to lifelong imprisonment for his views, Galileo was at least allowed to serve house arrest instead of prison time. In 1642 the man who Stephen Hawking and Albert Einstein both credited as the father of modern science died a heretic. It wasn't until 1992 that the Church, under Pope John Paul II, publicly acknowledged its wrongdoing.

Although Galileo may not have always received the credit he deserved, his work was not lost. Isaac Newton would bring Galileo's concept of inertia to life, turning it into the first of his three laws of motion, published in 1687 as part of his monumental work *Mathematical Principles of Natural Philosophy*. Only the third law was truly original to Newton. The first two laws summarized what others had already shown experimentally. As Newton noted, these laws were "accepted by mathematicians and confirmed by experiments of many kinds." With the first two laws, Newton wasn't setting out to break new ground. Instead, he sought to unify Galileo's physics and the astronomer Johannes Kepler's laws of planetary motion.[2]

The first law of motion states that objects at rest will remain so until they're acted on by an outside force. This essentially restates Galileo's principle of inertia, although you won't find the word *inertia* anywhere in *Principia*. Instead, Newton adopted Kepler's term: "innate forces." Eventually, this wording was replaced with *inertia,* and today, the first law of motion is also known as the law of inertia. With so-called innate forces acting on all matter, Newton wanted to define "force" mathematically. This would be his second law: the change in motion of an object is proportional to the force applied to it.[3] Today, the second law is expressed mathematically as $F=ma$. This equation, showing force as a product of the mass of an object times its acceleration, is fundamental to the study of physics and engineering. Some say it's the most important equation in physics, enabling us to quantify, predict, and determine relationships between forces and motion. Engineers use it to build bridges, design cars, and launch rockets— regardless of whether a physical system is at rest or in motion.

Unlike the first two laws of motion, Newton's third law does introduce a new concept. Everyone knows the third law of motion, whether they realize it or not. It's often expressed as "for every action, there is an equal and opposite reaction." The third law is also referred to as action-reaction. As Newton wrote in *Principia,* if you press your finger against a stone, the stone presses back with equal force. With this third law, Newton introduced a fundamental concept in physics: the conservation of momentum. Momentum is the measure of the motion of an object—specifically how much mass is moving and how fast it is moving. As a result of the third law of motion, if one object loses momentum, another must gain it.

Together, the three laws of motion offer insight into why friction is everywhere. Picture a book sitting on a table. We know from the first law that it will remain there until an outside force acts on it, such as a hand sliding it across the table. When this happens, the book and table experience friction. The book will exert a force of friction on the table and the table will reciprocate. It's up to our hand to provide enough force to overcome the friction and produce motion.

We can figure out how much force to apply to the book by employing the second law of motion, $F=ma$. Given how often friction is ignored in high school physics classes, one might assume that friction violates the laws of motion when, in fact, the opposite is true. Not only does friction not violate these laws—its existence explains

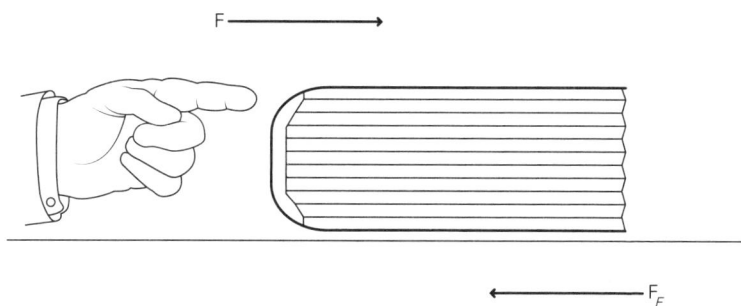

The force of friction acting against moving a book on a table.

them. Without friction, objects would move in perpetuity, which we know is not the case. Woe be to the engineer who designs a system and neglects to account for friction. The second law will be violated, the force calculations will be anything but correct, and system failure is a given.

Contact Mechanics

While growing up in Florida, I often encountered a small reptile called a green anole. Native to the southern United States, these reptiles play an important ecological role, controlling invertebrate populations. They're also territorial and feisty, which my cats learned the hard way. My cats would try to play with any green anoles that wandered inside, only to be outsmarted. They would yowl in annoyance when a green anole latched onto a paw. I became quite the expert at safely detaching anoles from squirming cats. Then, seemingly overnight, they stopped tormenting my cats and virtually disappeared. Instead, I noticed brown anoles, native to the Caribbean. A few arrived at first, and then the Floridian anole population turned brown as the native green anoles lost their fight against urbanization and a sturdier competing species. You can imagine my excitement when I returned home one recent summer and noticed something skittering along the boardwalk. A green anole—the first one I had seen in years.

My mom commented that she had started seeing them again—not abundant sightings like before, but sightings nonetheless. For the sake of survival, the green anole had learned to coexist with the brown anole by climbing higher in trees, above the brown anole's territory. This was possible through rapid evolution. The green anole's toes had grown larger and stickier. This created more friction, enabling green anoles to reach higher limbs. These changes occurred within about three years, an example of what biologists call character displacement, whereby species living in the same geographical region differentiate to minimize competition. Charles Darwin once studied the phenomenon in the Galapagos Islands. Observing differences in beak length among finches, he proposed that the beak of an ancestral species had evolved, enabling the finches to acquire different food sources. Like finches, the green anole had evolved—in this case, to use friction in its favor.[4]

Friction-friendly feet aren't just helping green anoles. Polar bears have also evolved to use friction to their advantage in an arctic climate. These magnificent animals adeptly navigate snow and ice thanks to dermal bumps on their paws that function like the treads on the soles of our shoes. We don't often associate anoles and polar bears, but both have adaptations that allow them to manipulate friction with their feet.

At this point, you might be wondering if adding bumps to all surfaces will increase friction and minimize slips. There is a widely held belief that rough surfaces cause friction. Like all generalizations, there is a "yes, but" component to this. Friction originates from the contact between surfaces, and roughness will influence that contact. But a tribologist will ask, "What is rough?" This is where the world of tribology collides with contact mechanics, a discipline that explores the mechanics of two bodies in contact with each other.

When you run your fingers along a surface, you might conclude it's either rough or smooth. All surfaces are rough, depending on the scale at which you study them. Even that seemingly polished mirror you just brushed your teeth in front of. Zoom in on the mirror, and you'll start to notice the hills and valleys that cover its surface. The high points of a surface are referred to as *asperities*. As two surfaces are pressed together, those are the points that make contact first. And where contact occurs, so too does friction.

The story of contact mechanics begins with an overachiever at Christmas break, 1880. My university breaks in December were usually spent visiting family and catching up on sleep. Fortunately for science, twenty-three-year-old Heinrich Hertz was a bit more ambitious. He developed the theory that would become the foundation of contact mechanics.[5] Earning his PhD earlier that year from the University of Berlin, Hertz continued working at the university as an assistant to Hermann von Helmholtz. Helmholtz is perhaps best known for his treatise on the conservation of energy, which explains why energy cannot be created or destroyed. But Helmholtz worked on a variety of problems related to optics, physiology, electricity, and philosophy of science. None of this may sound like contact mechanics to you, but optics is the link.

The Physical Society of Berlin had hosted demonstrations of a phenomenon known as Newton's rings, in which glass spheres pressed together create an interference pattern of alternating light and dark rings. At the time, scientists believed that the contact between rigid bodies would itself be rigid. Hertz, however, hypothesized that there might be some deformation at the point where the balls touched. Why else would you see that interference pattern dance around as the balls were pressed together? Hertz also proposed that the contact points of solid materials were elastic, meaning they would return to their original shape when no longer in contact.

To understand Hertz's proposal, we first need to distinguish between pressure and stress. *Pressure* is the amount of force applied to an object per area. *Stress* is the force per area that the material experiences. They sound similar, but one force, pressure, is an external application, and the other is the internal response. Hertz reasoned that if contact occurred at a single rigid point, as had previously been theorized, then the pressure applied at that point would essentially be infinite. This is because the area of a single rigid point would be so small. Since pressure is the force acting over an area, as area approaches zero, pressure approaches infinity. With pressure that high, there would be enough stress in the material to deform it, expanding the contact area from a small point to a larger area and distributing pressure and stress across it.

To prove this, Hertz made a key assumption: the contact area was small—very small. This meant that the shape of the objects wouldn't matter. By the time you zoomed in on the microscopic contact area, it wouldn't matter if the contact was part of a block, ball, or pyramid. Hertz also assumed that the contact area was elliptical, just like the observed rings. From there, he derived the equation for the pressure acting across the contact zone, and, finally, the resulting equation for deformation.

Understanding these high-pressure zones and how a material responds at the contact zone remain key aspects of design today, enabling engineers to design safe and efficient systems. Using contact mechanics and Hertz's equations, engineers can determine if the concentrated contact stress will give rise to a crack in the material, and whether the crack will propagate during operation, causing failure. Knowing that, they can assess if a material change will remedy the issue or if a part redesign is necessary. Engineers will do a quick check, using what is known as a Hertzian contact stress calculator. With a free online Hertzian contact stress calculator, you too can tinker around with the same problem Hertz tackled over a century ago.[6]

Hertz presented his findings to the Physical Society of Berlin and published them in the *Journal for Pure and Applied Mathematics*. At least his work would not suffer the same fate as da Vinci's! After publication, Hertz followed up his theoretical work with experiments to validate his findings. He coated one lens with a black dye and loaded it against another lens, enabling him to visualize the size and shape of the resulting contact area under a microscope. As he applied more load, the contact area grew, confirming that the surfaces were deforming. He then quantified the resulting contact area, showing that it grew proportionally to the load raised to the two-thirds power.

Friction originates at the contact between surfaces, and thanks to Hertz, we now know that the behavior of those contacts is dependent on the contact area. But in Chapter 1, we saw that friction is *independent* of contact area. This brings us to a crucial concept in contact mechanics—the distinction between real and apparent contact area. When sliding blocks across surfaces, Coulomb and Amontons measured the entire contact area of surfaces, not necessarily the actual area in contact. This is called the apparent contact area. Picture your shoe. It might appear that the entire sole is in contact with the ground, but actually only the treads are. Those parts are the real contact area. Hertz's model of contacts is concerned with the real contact area, the points where surfaces touch. He assumed an idealized surface of a single contact point. If you were to zoom out, the surface would look like a series of uniform half spheres. However, if you've observed a surface under a microscope, you know that a real surface is not a set of uniform and perfect half spheres. To understand friction and its relationship to real contact area, we need a more detailed picture of what surfaces in contact look like.

Enter James Greenwood and J. Bryan Williamson, two engineers who pioneered a statistical approach for describing how rough a surface is. Their 1966 publication describes one of the most widely used methods for quantifying rough surfaces. Its popularity stems from its elegance and ability to be "good enough," despite plenty of other methods developed over the following decades. Greenwood and Williamson met while studying and working in Frank Philip Bowden's laboratory at Cambridge University. Bowden was a tribologist, and you'll hear all about him later. Greenwood had studied mathematics at Clare College, Cambridge, and had joined the University Rambling Club. On a walk one day, a nail in someone's boot caused a spark, prompting a lively discussion with another member of the club, who was working in Bowden's laboratory. Months later, when Greenwood was pondering what to do after graduation, the hiking friend suggested that Greenwood join Bowden's lab group, where he met and worked with Williamson.[7] And so a sparking nail led to an important and productive collaboration.

Greenwood and Williamson were the first to digitally map surfaces. Using a profilometer, an instrument with a probe stylus that scans the surface of an object, they found that the heights of asperities follow a normal distribution with an average asperity height. This means that the probability of finding an asperity taller or shorter than the average height decreases the more extreme the height. Knowing this, Greenwood and Williamson could determine the number of asperities in contact and then, by treating each asperity as a Hertzian contact, sum up individual contact zones to calculate the real contact area of a surface. They found that with multiple asperities in contact, as the load increases, the total contact area grows in proportion to the applied load. An individual contact zone might grow at Hertz's two-thirds rate, but the sum of the total contact areas grows in an approximately linear fashion.[8] Amontons's laws of friction hold. At least, for the most part.

Atomic force microscopy (AFM) has confirmed these results by enabling precise measurements of friction at individual asperities. This high-resolution instrument consists of a small cantilever that scans surfaces at a sub-nanometer scale. It uses the atomic interactions between the probe tip and surface to move the cantilever up and down a surface, measuring roughness and friction between asperities. These experiments have shown that while the friction of an individual asperity doesn't always follow a linear relationship with load, when you average over many asperities, the relationship between friction and load becomes linear again.[9] This discrepancy highlights how the behavior of friction changes at different scales. Just as physicists work to unite Einstein's quantum mechanics and Newtonian mechanics, bridging the laws of friction at these different scales remains the holy grail of tribology.

Adhesion

Greenwood and Williamson's insights into what rough surfaces in contact look like enable us to start connecting surface contacts to mechanisms of friction. Friction mechanisms are broadly divided into two categories. The first is *deformation*. Deformation, often called Coulomb friction, is the force required to deform asperities enough to break them with a plowing motion. This motion, specifically the energy required for one asperity to climb over another, causes the asperities to deform, with one squishing the other as it climbs over it and becomes free to slide. The other mechanism, discovered over a century before Hertz published his seminal work on the behavior of elastic contacts, is *adhesion*. Adhesion is bonding that results from attractive forces between asperity contacts. Adhesive bonds resist motion when we try moving one or both surfaces.

John Theophilus Desaguliers is probably best known for his contributions to Freemasonry; he also created the first planetarium in England in the 1730s.[10] When the Edict of Nantes was revoked in 1685, his Huguenot family fled France. Legend has it that the infant John was smuggled across the English Channel in a barrel aboard a refugee ship.

The family settled near London, and John went on to study at the University of Oxford, where he became a natural philosopher.[11] Desaguliers had a gift for explaining physics to the public. As part of the first generation of Newtonians, he published and gave lectures defending Newtonian principles, including the laws of motion. In 1713 he relocated from Oxford back to London to give a series of public lectures on experimental philosophy, which is based on the belief that observation and experiments should guide scientific inquiry. Newton, the head of the Royal Society, took notice of Desaguliers, inviting him to give weekly demonstrations to the society. Soon after, Desaguliers was elected a fellow and tasked with conducting experiments that would validate Newtonian physics. While Desaguliers was not the only natural philosopher at the time to make a career of lecturing on Newtonian mechanics, he was one of the most effective, speaking to audiences in French, English, and Latin. He communicated concepts in an accessible way that the public could understand, much like a TED speaker today.[12] England was on the cusp of the Industrial Revolution, and there was such a thirst for physics that lecturers even held court with royalty. In fact, one of Desaguliers's experimental contraptions is part of an exhibit of the King George III collection.[13]

One of the principles that Desaguliers lectured on was gravity, a universal force of attraction between all bodies of matter. While demonstrating Newton's law of gravity and attractive forces, Desaguliers found that there could be an even stronger attractive force between objects than gravity. He demonstrated this with two small lead spheres, about an inch in diameter, by shaving off a small portion of each and, with a slight twisting motion, pressing them together. The result? It took more force for him to separate the spheres than it did to press them together. This, he deduced, was due to adhesion. (He called it "cohesion," which today refers to the interacting forces within a material.)[14] In subsequent lectures, he noted that adhesion between smooth surfaces could increase friction. It wasn't until the 1950s that two British physicists would combine both friction mechanisms, deformation and adhesion, into a single model of sliding friction—the one we use today.[15]

A Combined Mechanistic Model

Frank Philip Bowden and David Tabor are legends in tribology. Although they share a surname, Frank Philip, known simply as Philip, is not the Lord Bowden who commissioned the Jost report, which put tribology on the map. Philip was born in Tasmania, Australia, and became an eminent surface scientist. His lab in Cambridge is the very one where Greenwood and Williamson met. But David Tabor is the same David Tabor who consulted on the Jost report. He was affiliated with the tribophysics lab in Australia mentioned by the editor of the *Oxford English Dictionary* when suggesting the word *tribology*.

The two had quite the impact on tribology, despite Bowden's rather inauspicious start as a student. He originally failed to matriculate to university because he couldn't pass mathematics. Yes, the very same lab that the mathematician James Greenwood would one day join was started by someone who never felt very adept at the subject. But in a lesson of perseverance and how a weakness can become a strength, this aversion to math is precisely what made Bowden such an incredible scientist. He acknowledged that his accomplishments were the result of hard work, not necessarily the academic genius possessed by some, like Newton. Bowden chose paths that would avoid heavy lifting in math, such as theoretical physics and modeling, and instead pursued experimental science with a passion.[16]

While working as a lab assistant, Bowden continued his studies under a tutor and finally succeeded, matriculating at the University of Tasmania, where he pursued science. He eventually made the trek across the world to Cambridge University, at a time when the modern British PhD was still in its infancy. Many of us today probably can't fathom how a Cambridge PhD in the sciences could have initially failed high school math, but Bowden was not hindered by such a mindset and went on to head his own lab at Cambridge. There he would meet David Tabor, who in 1936 joined Bowden's lab for his PhD studies.

Bowden and Tabor soon began to suspect that the prevailing view of how friction arises wasn't quite complete. When Bowden and Tabor observed surfaces under a microscope, they noticed asperity junctions, points where attracting forces between the asperities become welded together and a force is required to shear the junctions apart.[17] This indicated that adhesive friction was occurring during sliding. Yet some junctions experienced plastic deformation, permanent deformation that occurs when a material surpasses the maximum stress that it can withstand. These asperities had been ploughed, releasing the contacts and enabling sliding. The total friction, they determined, was the sum of *both* mechanisms: deformation friction and adhesive friction. This discovery, known as the Bowden and Tabor model of sliding friction, resolved any questions around whether deformation or adhesion was the cause of friction. They both are.

The Bowden and Tabor model of sliding friction allows us to separate friction into its mechanistic components of adhesion and deformation friction. This isn't always of interest. Often, just the total system friction is what engineers are after. But if we're aiming to design a system to reduce or increase friction, breaking the friction into these components can help. Using properties of the materials involved, including hardness and yield strength, we can determine the value of the adhesive and deformation components of friction and, in turn, start designing for friction. But as our understanding of surfaces and friction has evolved, so too has Bowden and Tabor's model.

Adhesion and Deformation

By summing up the components of friction caused by the adhesion and deformation mechanisms, we treat them as having negligible interaction with each other. But this is not always the case, as Bowden and Tabor found. Even in adhesion, the asperities may temporarily deform, changing size and shape. And sometimes, permanent deformation can occur because of the breaking of adhesive bonds.

We'll start with the adhesive component of friction. As we saw, the asperities between contacting surfaces form junctions that sum up to the real contact area. These junctions can also form when the asperities are within close enough proximity for an attractive adhesive force to weld them together. This adhesive force may be chemical or physical, meaning that the surfaces are held together by chemical means—sharing or exchanging of electrons—or mechanically through physical bonding. This is in itself a source of scientific debate. Some would argue that bonds are physical, and others would argue that they're chemical. For our purposes, we'll assume that chemical bonds are intramolecular. They're formed as electrons are shared between atoms. Physical bonds are intermolecular, the result of interactions between atoms and molecules called van der Waals forces. These distance-dependent forces are relatively weak compared to intramolecular forces because they do not involve sharing or exchanging electrons. As electrons move around in the electron cloud surrounding the nucleus of the atom, their distribution can change briefly, inducing a temporary charge on the atom. If all the electrons are concentrated on one side of the electron cloud, for example, that side of the atom becomes more negatively charged. The extra charge may cause the atom to attract or repulse neighboring atoms and molecules. The resulting force is the van der Waals force.

Van der Waals forces enable animals such as the green anole to climb up trees and branches. Setae, or hairs, on the anole's feet form van der Waals forces with a surface, providing the adhesion to seemingly defy gravity. The more setae, the more molecules are available to form van der Waals forces. Because van der Waals forces act over short distances, they don't keep the anole permanently stuck to a tree. Anoles can peel their setae off and keep climbing by breaking the junctions between their setae and a surface.[18] This requires shearing the junctions. If you've ever played around with a suction cup, you know that the trick to separating it from a surface is to twist it, not pull it straight off. The twisting motion separates the surfaces, shearing the junctions. The ability of a material to resist this shearing is known as its *shear strength,* or the ability of a material to resist forces that cause it to slide or fracture along a plane.

As adhered surfaces slide over each other, shearing occurs parallel to the direction of sliding, just like friction. The resulting frictional force is related to the shear strength of the junction, which can be determined from the materials involved. However, shearing the interface does not mean you're shearing the bulk of the material. This is similar to how a fish can plunge into water, creating local ripples, while on the other side of the pond, the water remains completely still. The breaking of asperity junctions causes the equivalent of ripples and can lead to small, localized sections of the material being damaged and even removed. Broken adhesive bonds may not only lead to energy dissipation and thus friction, but may also wear away material, creating wear debris. That debris can create particles that then plow through the material, leading to deformation friction. And thus, one mechanism begets another.

Breaking and removing asperities is one way that adhesive friction can cause deformation friction. But there's another way. Tangential forces such as shear can impact the overall stress of a material, causing an asperity to yield and deform. In turn, the real contact area will increase. You can see this if you lay your hand on a glass surface, then gently press horizontally along the table. Your imprint will grow. This is called junction growth.[19] Deformation is occurring as a result of the attempt to break adhesive bonds. Now, the energy required to overcome friction is the energy needed to break the adhesive bonds and deform the material.

Even more puzzling, there's a specific type of adhesive mechanism called ratcheting that is almost analogous to Coulomb friction. Only in this case, permanent deformation doesn't occur. Ratcheting occurs when one surface contains narrower, sharper asperities than the other. During sliding, the smaller asperities climb up and down over the wider ones. Since the contact stresses are below yield, they don't plastically deform. However, this process, as you can imagine, impacts the force of friction. As the asperities climb, more energy is needed to shear the surfaces, giving rise to higher friction. When they descend, friction decreases, and less energy is needed. So to answer the question I posed earlier: Do we roughen surfaces to increase friction? Well, yes, surface roughness can clearly influence friction. However, as we'll see in Chapter 3, tribologists must be clever when engineering a surface because it can and will evolve over time.[20]

Revisiting Hertzian Contact with Adhesion

The Bowden and Tabor model of sliding friction seems almost too simple. Friction, the model indicates, stems from two mechanisms: deformation and adhesion. But there's a lot of gray area—these mechanisms are not always independent. This is common in tribology. Often, the simple, elegant model is good enough and generally describes the system. But we've just seen how much adhesion influences friction, and Hertz's model doesn't take adhesion into account. So, if we want to model the roughness of a surface to determine the total real contact area, do we just wave our hands?

Many times, we do wave our hands and stick to Hertz's model because it's "good enough." Science, however, doesn't stop at good enough. In 1971 three engineers published a paper on how adhesion influences the contact of elastic solids. Ken Johnson, an engineer, graduate student, and lecturer at the University of Manchester, was studying how to reduce vibrations in aircraft engines, which brought him into the realm of contacts and interfaces. Eventually his work caught the attention of David Tabor. During Tabor's visit to Manchester, the two met, and Tabor was impressed with Johnson, who was a meticulous experimentalist. Yet another lifelong friendship and academic partnership would transpire. Johnson's book on contact mechanics remains the authoritative text on the subject.[21]

Johnson joined the Cambridge faculty in 1954, not long after Bowden and Tabor published their theory of sliding friction. With his background in experimental contact mechanics, Johnson was an asset to the group, working on friction, contact mechanics, and, notably, adhesion. Kevin Kendall and Alan Roberts were two of Tabor's research students. Roberts was studying the friction and lubrication of windshield wipers to improve their performance. He had noticed that when he applied a load to a rubber windshield wiper and then removed it, the windshield wiper didn't fully separate from the glass beneath it. In fact, he had to apply an additional force to pull the wiper off the glass. Kendall suggested that the lingering contact might be related to the surface energy of the glass. This interaction would lead to adhesion, which was more pronounced with softer materials. By measuring the surface energy of two objects, Kendall and Roberts found that even lightly loaded objects have contact area growth from adhesion, which makes it harder to pry them apart. This is especially true of soft materials that we consider sticky, like rubber or gelatin. Such materials experience greater deformation and are therefore more likely to form adhesive bonds. Hard materials, however, like metal or ceramics, experience less contact area growth. Kendall and Roberts concluded that adhesion was pulling more material into contact than predicted by Hertzian contact mechanics. Now they needed to sort out why. Johnson was brought in to try to modify Hertz's theory to account for adhesion.[22]

Ken Johnson solved the problem the very next day while watching the Football Association Cup Final. His solution assumes a spherical contact. But in addition to the force that is being applied to the sphere, it considers forces that might be present between the surfaces themselves. Of particular interest are the areas of the sphere close to the contact, where van der Waals forces, if strong enough, can bring the surfaces into contact, increasing the contact area. When this happens, the attractive forces between the surfaces require a stronger force to separate them, as Roberts suggested. The Johnson model amends the Hertzian contact equations by including an adhesive term that accounts for van der Waals forces. The modified equation, known as the Johnson-Kendall-Roberts (JKR) model, describes how surfaces adhere together and the resulting contact area between them. It remains the most popular model for determining the contact behavior of soft materials.

As technology has evolved, so too has our understanding of contact mechanics. After JKR, other theories, such as the Derjaguin-Muller-Toporov (DMT) model, have been developed to predict the contact behavior of harder materials with minimal adhesive forces. Whereas the JKR model includes the long-range attractive forces between the materials in the contact zone, the DMT model ignores them. The development of AFM has been key to developing models for materials with different properties. In the late 1990s, Rob Carpick and his collaborators at Lawrence Berkeley National Laboratory presented an equation for materials that fall in between the more adhesive materials modeled well by JKR and the minimally adhesive materials modeled best by DMT.[23]

Bo Persson, a Swedish researcher at the Peter Grünberg Institute of Forschungszentrum Jülich, has even used fractal theory to model the contact area growth of rubber surfaces. In his model, surface roughness follows a fractal-like pattern, a geometric shape that repeats at various scales. Familiar examples include broccoli, pinecones, and succulents. If you peel off a piece of broccoli and examine it up close, you'll notice that it looks nearly identical to the original piece—albeit smaller. Persson showed that most surfaces are fractals. That enabled him to propose a radically new way of looking at surfaces in contact. Instead of viewing surfaces as asperities, his fractal model suggests that a surface looks the same at any location, as long as you're viewing it at the same magnification. The way a surface responds to applied forces and pressure depends on the level of magnification. Zoom out of the surfaces in contact, and it will appear that the entire surface is in contact. But zoom in and you'll see that the zones without contact grow. This means the pressure distribution and magnitude of the stress within the material vary across scales, affecting the material's response and the resulting contact area.[24] Persson's theory produces a more accurate calculation of real contact area. This comes at a cost, though, as it involves more complex mathematics than other theories. As a result, the Hertz and JKR models remain dominant. They're not only simple and elegant but usually good enough.[25]

The Hertz and JKR equations have been used in evolutionary biology to explain how critters such as the green anole, but also spiders and flies, can climb vertical structures. Insects have setae diameters that are on the order of micrometers. Geckos, lizards, and anoles have even thinner, sub-micron setae, despite being larger than insects. This may seem counterintuitive, but it can be explained by contact mechanics. Larger animals, like geckos, split up their contacting setae to increase the number of contact zones, creating more places for adhesion via van der Waals forces to occur.

With this knowledge, engineers are at work designing adhesive systems that mimic those we see in nature. In particular, researchers have been trying to develop materials that mimic the adhesion of geckos since a gecko-like adhesive could stick to surfaces without any special preparation. When you hang a hook in a college dorm room, you must clean the wall before mounting if you want it to stick. But geckos can climb up anything. Gecko-like adhesives could be used in manufacturing to pick up and move variously sized items from production lines without damaging them. They could join electronic components at lower temperatures, reducing the energy costs of manufacturing. And they could be used to make bandages that are easier to apply and peel off. Nature, however, always seems to outperform what is made in the lab.[26]

Creating a sticky surface isn't the only way to imitate adhesion in the natural world. Another way is through engineering design. The ability to adhere to surfaces evolved independently in anoles and geckos. Anoles have miniscule pads with higher density, unbranched fibrils that are shorter and have one tip per fibril. Geckos have longer, more complex branched fibrils with hundreds of tips. This design difference may explain differences between how anoles and geckos detach themselves from surfaces as they move. Geckos peel their toes from the outer edge to the inner edge; anoles peel them in the reverse order, from inner to outer edges.[27] Despite these differences, the adhesive performance of an anole and a gecko has been found to be similar. Using the JKR model, researchers have calculated the force that geckos and anoles use to pull their feet off a surface. The anole may have a simpler structure, but thanks to the very high density of its fibrils, it's able to create enough contact area to stick to surfaces with as much strength as a gecko. These insights into the mechanics of how the gecko and anole adhesive systems work may be key to helping researchers finally mimic their adhesive magic through engineering design.

Friction and Materials

Geckos and anoles might look similar, but they have evolved distinct ways of manipulating friction. This highlights how sensitive friction is to small differences between systems. You can imagine, then, how much friction behavior varies across vastly different materials and objects. There's a reason we classify materials broadly into metals, polymers, and ceramics. Each have different structures, different properties, different applications, different performance, and different amounts of friction.

Machine components are overwhelmingly made of metals. Metals—copper, aluminum, iron and steel, and their alloys—are strong, durable, and maintain their mechanical and chemical properties at extreme temperatures. These properties stem from the bonding that occurs between the atoms of metals. In metals, the outermost electrons move freely around the ordered arrangement of atoms, known as a crystal structure. These delocalized electrons form strong attractive forces with the positive nuclei of the atoms, giving metals their strength. This is called *metallic bonding*. These freely moving electrons explain why metals are conductive and able to be deformed without losing their strength.

In an ideal world, we would have a nice little table that listed the coefficient of friction for specific metals. However, friction is a system property that depends on every condition present, including the other materials in the system. If someone tries to sell you on buying a material because it has a specific coefficient of friction, you are now completely in the right to call shenanigans on them. There are, however, tables with coefficients of friction for metal pairs, like copper on copper or copper on steel. These tables usually have a range of values and specify the conditions for which the values apply. But buyer beware. If your conditions aren't identical, you may experience drastically different friction—so drastic that it could tear up your equipment and lead to failure.

Clean, metallic surfaces can have really high friction. The coefficient of friction can even exceed one, which sets off alarm bells for tribologists, who aim for a much lower coefficient. Throw those metals into a high vacuum and watch the coefficient of friction climb even higher— up to ten. In a vacuum, any gas or liquid molecules that may have attached to the surface of a material are pulled off. Without contaminants, junction growth between metal contacts is unhindered and becomes excessive, leading friction to skyrocket and the parts to seize up. This doesn't mean we can't slide metallic parts in vacuum environments. It just means we have to engineer around this problem, as we'll see in Chapter 3.[28]

Another aspect of metallic friction that can be problematic is something the Romans were very familiar with: frictional heating. When surfaces slide over each other, the energy released due to friction is concentrated over small areas. This causes bursts of high energy at the contacts between local asperities. The frictional energy becomes thermal energy, or heat, and leads to flashes of high temperature in the contact zones. The bulk of the object may remain mostly unaffected, but the contacting asperities can heat up to the point of melting, depending on the temperature limits of the material. Metal parts can get hot enough to discolor steel disks. As you can imagine, this is not ideal. Just as the Roman chariot racers didn't want hot feet, you wouldn't want to risk burning yourself on these frictional pairs. Equally important, nearby materials must be able to handle the heat. No one wants a melted gear shutting down production.

If frictional heating is not managed properly, it can alter metallic systems, triggering chemical reactions, material property changes, and phase transitions. Chemical reactions such as oxidation, the process that causes iron to rust when its molecules react with oxygen, can change the surface roughness profile of a metal by creating a new layer of the oxidized material, affecting friction. Temperature can also affect the mechanical properties of materials, such as their strength, ductility, and elasticity. These properties determine whether materials deform plastically or elastically in response to pressure and stress. Finally, if you heat a metal, at a certain point it can undergo phase transitions that impact friction. This effect is well illustrated by cobalt. When heated, its atoms transition from forming a hexagonal structure to a very ductile cubic structure, with atoms on each corner and face of the cube. When this happens, its coefficient of friction increases significantly.

If you're looking for materials that are stable at high temperatures, or more stable in a range of environments, ceramics may be what you're after. Ceramics, which include porcelain, stone, concrete, and some specially engineered materials, boast high mechanical strength and resistance to corrosion. Plus, they perform well at high temperatures. This combination of properties is why ceramics are used for cookware, medical implants, and even protective heat shields on planes and spacecraft. If this sounds too good to be true, here's the catch: ceramics are brittle. Fracture toughness, the fancy way of describing how well a material resists cracking, is extremely important for ceramics. But they'll never be as ductile as metals. The bonds between atoms are much stronger than those of metals, which makes it harder for their atoms to move past each other. Like metals, ceramics form crystalline structures. However, the atoms are held together by ionic and covalent bonds. Ionic bonds occur between metal and nonmetal atoms. The electron of the metal is transferred to the nonmetal, creating a positively charged metal ion and a negatively charged nonmetal ion. Having opposite charges, the two ions are strongly attracted to each other. In a covalent bond, an electron pair is shared between two nonmetal atoms. These strong bonds hold the atoms tightly, leading to their brittleness. Ceramics crack rather than deform and recover. Some ceramics, like diamond, have microstructures that actually *do* allow movement between planes of atoms but require incredibly high applied stress to do so.

Because they are so rigid, the real contact area for ceramics is low compared to a pair of metal surfaces. In general, that means ceramics in contact will likely have lower friction than clean metals in a vacuum. That's not to say all ceramics have low friction. Despite their reputation for being robust against chemical attack, they aren't completely immune to their environment. The friction of ceramics is another one for the "It depends" folder. The depths of that folder will be explored in Chapter 3, where we delve into the tricks tribologists use to manipulate friction.

While I have tinkered with some ceramics, the majority of my work has involved polymer materials running against metals. "Polymer" is the fancier term for plastics and elastomers, which themselves are fancier terms for materials like rubber and putty. I love polymers, but they can be difficult to work with. Based on your own daily interactions with the materials around you, you probably already know that polymers tend to be weaker than metals and ceramics. But they're also more flexible and lightweight. A polymer has a long backbone, mostly composed of carbon atoms, from which smaller chains branch. The atoms in the backbone are covalently bonded, and the chains are held together by van der Waals forces. These relatively weak forces make the chains more flexible than ceramics and metals. Because there's such a variety of backbones and side groups, polymers exhibit a range of properties. They can be flexible like putty or stiff like the polyimides used in jet engines. If you compare the friction of polymers to metals and ceramics, you'll find that friction involving polymers is lower than that of ceramics and metals because of polymers' weaker bonds and greater flexibility. This is largely true, but you can also get pretty high friction from polymers. Rubbers, for example, are a class of polymers that we often use specifically for their high friction, such as on the soles of our shoes so we don't slip and fall.

Perhaps the most striking difference between metals, ceramics, and polymers is their contact behavior. Polymers in contact, either with another polymer or a metal, typically form elastic contacts, meaning deformation will cease once the loading and stress being applied are removed. Metals on metals form more plastic contacts, which result in permanent deformation. Tribologists use a plasticity index, developed by none other than Greenwood and Williamson, to determine how asperities will behave under an applied load and what friction mechanisms they can expect. The plasticity index indicates whether asperities will experience elastic flow, in which the material deforms under an applied force but recovers its original shape once the force is removed, or plastic flow, in which the material is permanently deformed. The plasticity index is based on the elasticity, hardness, and roughness of the surface. A plasticity index greater than one indicates plastic flow will occur. Once the index drops below one, the flow becomes elastic. Polymers are, in very scientific terms, squishier than ceramics and metals, so they more easily deform elastically when in contact. Their plasticity index is about one-tenth of the plasticity index of metals or ceramics. This means that polymer contacts are largely elastic, and adhesion is often the dominant mechanism of polymer friction.[29]

The other important difference between polymers and metals and ceramics is that almost all polymers are viscoelastic. Viscoelasticity describes a material that exhibits viscous and elastic behavior as it's deformed. Elasticity is the solid counterpart to that: resistance to deformation. Viscosity is a liquid's ability to resist flow. And so a viscoelastic material acts a little bit like a solid and a little bit like a liquid. If you have ever pulled Silly Putty, you know what this seemingly contradictory description means. You can quickly rip it apart as if it's a solid, but when you stretch it slowly, it behaves almost like a liquid.

A key aspect of viscoelastic materials is that they experience time-dependent properties. When you push against or deform a viscoelastic material, there's a delayed response. Once you let go, it slowly relaxes. This behavior arises because polymers are made up of long chains. As you apply stress, the chains become tangled, and the behavior is more elastic and the material more like a solid. Releasing the stress enables the polymers to untangle, flow easier, and eventually return to their original shape. For example, if you set Silly Putty on a table and leave it for a few days, it will spread out to cover a much wider area. Owing to this time-dependent behavior, the contact areas of viscoelastic materials may not follow the same linear trends as other materials. This means that sometimes polymers go rogue and break some of the rules of friction. And so polymer friction can be expressed in two words: it's complicated.

Static Friction

Until now, we've mostly discussed the friction of sliding surfaces but not how they *start* sliding. Static friction, the frictional force that must be overcome for movement to commence, is a time-dependent behavior. Once again, we have Coulomb to thank for advancing our understanding of this force. He observed that the friction between a wood sample loaded on an iron bed multiplied after being in contact over four days.[30] The startup friction, if you will, for the system was higher than when he set up the experiment and ran it immediately. The longer a wood sample is loaded on an iron bed, the more opportunity the atoms have to interact. When bonds are mated, more force is required to enable motion, increasing the static friction. If you've ever lightly laid a piece of tape on a table and come back later to find it completely stuck to the surface, that's an example of static friction. The opposite can also happen, particularly with a "dirty" surface. A contaminant can enter the picture, reducing friction, like a banana peel between your foot and the floor. (Why a banana peel? I can't recall ever seeing someone slip on a banana peel, but you get the point.)

It's easy to see why static friction is viewed as a problem. When you turn the key in your car ignition, you want the car to start and not stall because of static friction. But static friction isn't the enemy in every situation. In fact, static friction is not only necessary but critical when it comes to Earth's plate tectonics.

It's safe to say we don't expect Earth to spontaneously open and then close up again. But in 2017, researchers at the California Institute of Technology (CalTech) and the École normale supérieure in Paris showed that it can happen during an earthquake.[31] Earth is covered in faults, areas of fractures between two slabs of rock. These fractures enable the slabs to move, causing earthquakes. In a normal fault, the slab of earth located above the fault moves downward. In thrust faults, the opposite occurs. The lower slab pushes against the upper slab with enough force to overcome the static friction keeping them stationary, moving the upper slab up and over the lower one. Thrust faults have been responsible for major earthquakes, including the 2011 earthquake in Tohoku, Japan, that claimed the lives of nearly twenty thousand people. It had long been assumed that at shallow depths, this sliding motion could occur only over small distances, and it would therefore be impossible for the earth to open up. But after studying the Tohoku earthquake, the researchers found that the fault had slipped nearly fifty meters, enabling the tsunami that devastated the Fukushima Daiichi Power Plant.

To figure out how such extensive slipping could occur, the researchers recreated the earthquake, using plastic blocks that had the same mechanical properties as slabs of rock, and placed a wire fuse where the epicenter of the earthquake was. After assembling the plastic blocks under the tectonic load of the fault line, they ignited the fuse, triggering the earthquake. The friction between the blocks at the fuse location dropped sharply, enabling the rupture to propagate quickly down the fault. The researchers had chosen transparent and photoelastic blocks, whose optical properties change under different applied forces, to visualize the stress waves as they moved through the material during the experiment. What they found was unexpected: the fault momentarily twisted open and then snapped shut. Not only did this twisting motion explain the large amount of slip, but it provided insight into how tsunamis are generated. As the blocks ruptured and slipped along the ocean floor, their movement set the water in motion, generating the tsunami. All this was caused by the reduction in static friction holding the plates together. When it comes to thrust faults, we should all root for static friction.[32]

If we want things to remain in place, whether they're slabs of earth or more benign objects like the nail in the wall holding up your vacation photo, we need static friction. When we want them to move, however, we want to minimize it. Anyone who has had to push heavy furniture around can attest to this. Unfortunately, static friction sometimes sneaks in when objects are already in motion—when they're oscillating or sliding back and forth. That's because at the instant an object changes direction, there's essentially no motion. The object has to start up again in the opposite direction, which means it has to overcome static friction.

Static friction is responsible for a phenomenon called *stick slip*. Sometimes, sliding surfaces get stuck, and after overcoming static friction again, one of them slips. This most often happens after a reversal in direction, but it also happens during sliding. Stick slip is the result of the frictional force not being constant during sliding, which can occur for a variety of reasons, such as changing surfaces, wear, or environmental conditions. It's sometimes referred to as chatter since the resulting plot of friction versus time looks like a sawtooth when measured. Friction can increase when the object encounters deformation or adhesion of asperities and decrease as the slip occurs, creating a chattering signal on a plot.

Stick slip isn't always a bad thing. We have found ways to work with it and even use it to our advantage. The stick slip on a cello or a violin can be downright lovely. I say *can* because if you listened to me learning the viola, you know that even a gorgeous stringed instrument in the wrong hands can make stick slip sound horrible. In the right hands, however, the stick slip of a stringed instrument might lead you to believe all stick slip sounds glorious. However, stick slip can also be noisy since it produces vibrations, creating the shiver-inducing sound of chalk scraping across a board or the squealing of a brake instead of a violin concerto.

When you talk to a violinist (or cellist or violist), they'll tell you their bow is just as important to them as the instrument itself. Bows are constructed from a variety of woods, even plastic and composites in some cases, and different types of hair. Natural horsehair is the preference of musicians who want fine control and whose ears can detect the slightest change in friction between their strings and the bow. Neophytes like me tend to plonk around with synthetic fibers.

When researchers at the Madrid Institute of Advanced Studies in Nanoscience used AFM to examine how structural differences between natural and synthetic bow hair affect sound, they found that the roughness profile of the bow fiber is critical to better audio quality and finer control. Two aspects of the roughness contribute to the audio quality: how frequently the higher points of roughness on the surface occur and the roughness scale of the fiber compared to the instrument string. When the roughness peaks and scale were similar to that of the string, the audio quality was superior. Applying rosin to the bow also helped improve the stick slip, yielding similar audio quality between synthetic and natural bows. The problem with rosin is that it can be messy. It needs to be reapplied regularly. It can also get too sticky and must be replaced. The results indicate that with designed surface structure, stick slip could be better controlled without the need for rosin. This work can help not only with modeling and predicting stick slip. It might ultimately improve the design of synthetic bow hair to produce audio quality on par with that of natural bow fiber.[33]

Modifying the roughness of a surface can help manage stick slip. However, since there's more than one type of stick slip, there's no panacea for it. In the case of clean, smooth metallic contacts, stick slip is caused by adhesion between asperities. If a viscoelastic material is involved, stick slip may arise from structural changes in the material as it experiences different stresses during operation.

In my work, I've yet to design intentionally to make stick slip happen, but never say never. I have, on the other hand, spent plenty of time trying to design systems to eliminate stick slip. That was crucial to the vaccine syringe I worked on. Just imagine the stopper chattering down the barrel while trying to convince a child that getting a vaccine isn't so bad. A lot of tribology goes into making a vaccine that can be administered smoothly. So, while a violinist might want to harness the magic of stick slip, you can see why, in other cases, stick slip is not what the doctor ordered.

Rolling Friction

In Chapter 1, we saw that it's easier to roll an object than slide it. Rolling friction has a much lower coefficient of friction than sliding friction because of the contact a roller creates with a surface. During sliding, a lot of area comes into contact. During rolling, the contact area is orders of magnitude lower. With fewer contact points, there are fewer places for friction to occur, leading to overall lower friction. Friction from rolling is still composed of deformation and adhesion components. Whatever adhesion does occur during rolling is broken apart as the roller rotates out of contact. Rolling produces tension, or lifting, which pulls the bonds apart as opposed to shearing them in the plane of motion. While adhesion can be a significant contributor to sliding friction, it can be much less significant during rolling because of these different mechanics. It's also worth noting that surfaces move in and out of contact often as one surface rolls over another. This can result in plastic deformation of the surfaces over time. This plastic deformation may lead to more contact area and increased friction, which may or may not be desirable. To control rolling friction, tribologists manipulate the contact area through engineering design and material properties. Various factors can affect rolling friction: load, velocity, and as anyone who drives knows, environmental conditions such as temperature and moisture.

Throughout this book, you'll notice that tires will keep coming up as an example of why we need rolling friction and why we often want to optimize it. Tires are an engineering feat. To achieve enough traction to propel our vehicles forward, we want to maximize the contact area between the tire and the road. This is accomplished with a smooth tire without treads, the kind of tire used in NASCAR racing. When it rains and the track is wet, however, NASCAR races must be postponed. Smooth tires slide all over the place if a layer of water forms between the tire and the road. But F1 tires can be used in rain, as long as it's not a torrential downpour. F1 tires have treads that displace water. Treads are the magic in tires that mitigate the friction challenges that arise in different driving conditions.

As a car owner, I regularly glance at the treads on the tires. If the tread gets too low, I know it's time to replace the tire or risk not having good control of the car on the road. But treads decrease the contact area, so how do they help create traction, which stems from friction? Tires need to have friction in both wet and dry conditions, a headache of a problem since fluids bring about a whole new set of physics to contend with. If you've shopped for tires, you've probably noticed that the treads on all-weather and winter tires are much more intense than those on other tires. Those treads provide channels through which water can travel, enabling the higher surfaces of the treads to remain in contact with the road. So while treads decrease the contact area in dry conditions, they significantly increase it in wet conditions.

Otherwise, the water would create a film under the tire and increase the chance that the car will hydroplane.

Temperature also affects tire friction. The elastomers used in tires have different stiffnesses in hot versus cold weather. In hot weather, they become more pliant, and the real contact area approaches the apparent contact area, increasing traction. When it's cold, they stiffen, reducing contact area and traction with the road. Special compounds are used in winter tires to counteract this. At the end of the day, engineers must balance a lot of variables to ensure that tires achieve the right amount of friction with the road. We only seem to notice tires when something goes wrong . . . usually when we fail to maintain them. But that's friction, isn't it? Unseen and unappreciated until we're forced to contend with it.

Breaking the Laws

When we need friction, as we often do, it would seem obvious to turn back to the laws of friction to figure out how to increase it. We know from the first law that friction is proportional to load and that if we increase the load, friction will increase. The other two laws state that friction is independent of apparent contact area and sliding velocity, so we can forget about using size and speed to increase friction. Or can we?

Unlike many laws in science, the laws of friction can be broken at different scales and under different conditions. We already discussed the controversy around the relationship between load and contact area when we discussed modeling surfaces and contact mechanics. Even Coulomb knew from his experiments that the law we now credit to him, the third one regarding velocity independence, had exceptions.

One of the quickest ways to throw out these rules is to add fluid lubrication. That's because drag, or fluid friction, is dependent on velocity. Things get significantly more complicated when fluids enter the picture, which is why you'll get an entire chapter devoted to them. A second way to encounter exceptions is to roll an object instead of sliding it; Coulomb's laws pertain to sliding friction. Dry sliding friction, to be more precise.

Another way to assign an asterisk to the rules of friction is to introduce polymers. Metals and ceramics largely obey the laws of friction. But polymers have such a wide variety of material properties that they often go completely rogue. Under certain conditions, like low applied loads and surfaces that are neither very rough nor very polished, the real contact area is proportional to load, and friction remains constant. Check marks for laws one and two, then. It all goes awry when we start dealing with higher loads. As loads start to increase, the elastic contacts of polymers cause the coefficient of friction to decrease. Even though friction remains proportional to the real contact area, as the load increases, the real contact area is no longer proportional to the applied load.

With polymers, the elastic contacts can grow until they essentially are welded together, creating one large asperity instead of many asperity contacts. As we know from Hertz's work, the real contact area of a single asperity is proportional to the load raised to the two-thirds power. When we work through the Fun Friction equation, expressing F in terms of load, the coefficient of friction becomes inversely proportional to the load raised to the one-third power, and hence, less force is required for movement. Yet proceed with caution! For really soft polymers, merging asperities can cause the real contact area to approach the apparent contact area. As this happens, we return to Amontons's law, and friction is once again proportional to load, resulting in higher friction. More area in contact, more friction.[34]

At higher temperatures, polymers soften, enabling the molecular chains to move past each other. With continued heating, polymers can melt. Once the chains are mobile, the increase in sliding speed results in even more chain mobility, and a molten layer can appear between the contacts, reducing the friction in the system. This means that the third law of friction will no longer hold. Due to the wide range of polymer properties, such as stiffness, elasticity, and viscoelasticity, the relationship between temperature, friction, and sliding speed gets complicated and varies considerably among polymers. What we know with certainty is that temperature affects the friction of polymers.[35]

And this isn't just the case with polymers. The third law can be broken with other materials, including oxidized metals. Some metals readily oxidize in air or in the presence of other oxidizing agents. In this case, if the sliding conditions are aggressive enough, either because of high velocity, load, or a combination of the two, the oxidized surface will break up. The cycle of forming an oxide layer that is destroyed shortly thereafter leads to friction fluctuations. When this happens, friction is no longer a constant function of load or velocity.

We could spend this entire book digging into exceptions to the laws of friction. Some may argue that there are four or five laws of friction and that these laws address surface roughness and temperature as well as static and dynamic friction. But we'll stick with the three classical rules of friction that are well accepted. The best way to view friction's laws is that they hold until they go head-to-head with the laws of nature, which always take precedence.

The Origins of Friction

Humanity's relationship with friction was born of necessity, whether it was because we needed to create fire or make a wheel turn. But today, thanks to our ever-expanding knowledge of how friction originates, we don't just deal with it—we proactively design with it and for it. Despite how far we've come, the exceptions demonstrate the need for tribologists to continually study friction at different scales and under different conditions. This may entail challenging even the classic understanding of friction, such as the sliding friction model.

For example, in some metallic sliding systems, friction is so high that energy dissipation cannot be accounted for by the plastic deformation of asperities alone.[36] The wear rate would be significantly higher than what is observed. Some tribologists believe that energy dissipation can be explained by surface atoms, which aren't rigidly bonded but instead create phonons, packets of vibrational energy generated during sliding. In some cases, however, the systems are not well suited for phonon generation because of the arrangement of their atoms. This is believed to be due to the type of contact between the atoms of contacting surfaces. If contacts are commensurate, the spacing of their surface atoms is identical, and they align perfectly. Tribologists have found that during sliding, the oscillations are greater when the surface atoms don't align perfectly, or are incommensurate.[37] A working hypothesis is that commensurate contacts enable phonons to more easily propagate through a surface and thus contribute more to friction. Yet friction still arises between incommensurate contacts. Atomic-level friction is an ever-evolving field, at once challenging and illuminating. Connecting the behavior of friction at the micro and macro scales is getting tantalizingly closer as technology continues to advance. Even data science and machine learning are being employed to help resolve some of the ongoing gaps in our understanding of friction mechanisms at different scales.[38]

There are no easy answers when it comes to manipulating friction, unless that answer is "It depends." If you need to increase the friction in a system, you have to look at the system as a whole and determine the path forward that best suits the problem at hand. In the case of the survival of the green anole, their solution was increasing the real contact area by splitting up their setae. This increased adhesive friction enabled them to shimmy higher up a tree than their competitors. Nature evolved to increase friction, and so do our engineering designs. The tread of tires and the surface profiles and material properties of brakes maximize friction so that our cars grip the road and stop when they must. Even simple inclined planes are designed to increase friction when necessary. Nonslip ramps, with a rough surface and sticky material overlaid on a smoother and slipperier one, have saved me from wiping out on more than one occasion. All these designs stem from asperity interactions investigated in the twentieth century by the likes of Bowden and Tabor, and so many others.

For me, the names Bowden and Tabor are legendary. Many tribologists are able to trace their academic lineage back to Bowden's lab, which Tabor ran until the 1980s following Bowden's death in 1968. Greenwood and Williamson are only two names in an impressive cohort of alumni to have trained in their lab. You could say that Bowden and Tabor hold celebrity status within the tribological community. Tabor was described as gentle and kind. Tabor described Bowden as "humane" as well as charming and multitalented. And despite their great achievements, both had a reputation for humility. While the thought of devoting a lifetime to understanding friction might at first seem baffling, when we see the myriad ways this force affects us, we can appreciate the dedication of Bowden, Tabor, and every other tribologist. We need friction. From the movement of cells in our bodies to our work commute, it's there and it's necessary. Digging into the minutiae of contact mechanics and intermolecular forces helps us understand how anoles climb, how to warm our hands or make a fire, and how to stop things like our cars when we need motion to cease.

But friction is not always welcome. That boot spark that prompted Greenwood's tribology career? That's caused by static electricity, which can occur when charged atoms are transferred between surfaces as they rub against each other. It can be used to our advantage for electric braking, but we don't want to get statically charged simply by walking on carpet on a dry winter's day. As important as friction is, sometimes we need to do more than design with it in mind. We need to control and minimize it.

3 *When You Rub the Wrong Way*

NINETEEN MILES SOUTHEAST of Rome lies Lake Nemi, a volcanic crater lake about a mile wide. While picturesque, Lake Nemi would likely be little known outside the Lazio region of Italy, were it not for its archaeological significance. Legend had it that a gigantic ship lay at the bottom of the lake. This was curious, given that Lake Nemi is a tiny, landlocked body of water. In the fifteenth century, Prospero Colonna, nephew of the Pope, decided to satisfy his curiosity about the legend and sail out into the lake. After spotting wooden beams at the bottom, he and his men tried to retrieve the beams by sinking ropes with hooks attached to them. As you can imagine, they weren't terribly successful, but they managed to rip up some planks and confirm that something lay down there. This started a trend of fishermen sailing out and casting their nets to see what artifacts they could reel in. It wasn't until the 1890s that divers made their way to the bottom of the lake and observed the entire wreck.

Or, as it turns out, wrecks. A year after discovering the first ship, divers discovered a second ship. Soon, the fascist dictator Benito Mussolini took an interest in the site. In 1927 he demanded that the lake be drained, ordering his engineers to pump the water into an ancient Roman cistern that connected to surrounding farmland. Sixty-five feet of water were drained before one of the ships peeked out from the mud in 1929. A second ship was revealed in 1931. The largest ancient ships ever recovered, one measured 79 feet wide and 240 feet long, and the other 66 feet wide and 230 feet long. They were likely built sometime between AD 37 and AD 41 by the Roman emperor Caligula, a brutal leader who enjoyed ambitious construction projects, particularly when the output was luxury for himself. The purpose of these giant ships in a small, landlocked lake? Well, they were floating pleasure palaces. Ridiculous, perhaps, but their discovery disproved the idea that the ancient Romans couldn't build large ships. They also provided the earliest known evidence for the ball bearing, a rolling element used to ease the movement of mechanical parts and objects.[1]

The ships' wooden platforms were designed to rotate on a mechanical device. Eight bronze balls were mounted to the bottom of each platform and secured by a rod, enabling the platform to support and rotate heavy loads while reducing friction through the act of rolling.[2] This design was remarkably similar to that of the modern ball bearing.

Innovations around this clever invention remained stagnant until the Renaissance. It was then that the founding father of tribology, Leonardo da Vinci, was toiling with friction and came to study the ball bearing. We already know that da Vinci used pulleys to minimize friction. His sketches have also revealed that he developed the modern ball bearing to reduce the friction between contacting plates in his helicopter design.[3] Sketches made around the year 1500 depict eight balls contained in an octagonal raceway. One plate rests on top of the device, rotating with low friction, thanks to rolling provided by the ball bearings; the other plate does the same on the bottom. Da Vinci understood the importance of the spacing between the balls. While you may be inclined to pack as many balls into the raceway as possible, too many balls touching each other can hinder motion. Spacing them out allows for easy motion and the lowest possible friction.[4] Today, we have significantly improved manufacturing techniques to allow for precision designs. Yet the overall concept of modern roller bearings is nearly identical to what da Vinci designed at the turn of the sixteenth century.

As early as the 1660s, the polymath and physicist Robert Hooke noted that adding ball bearings between the wheels and axles of a carriage could be advantageous. Then, in 1734, the engineer Jacob Rowe designed and filed the first patent for roller bearings, which he called "friction wheels." These consisted of metal ball bearings placed between the wheel and axle to reduce friction in carts, wagons, and mills. Rowe realized, nearly two centuries before Jost, that reducing friction would have significant economic benefits. He calculated that if forty thousand horses were employed in the United Kingdom, then his friction wheels could reduce the number of horses by half, saving £1,500 daily in direct labor costs and almost £550,000 annually. This figure didn't include the cost of keeping horses; with that factored in, there was an additional £200,000 of savings available.[5]

Advances in manufacturing techniques during the Industrial Revolution made steel affordable through mass production. Ball bearings could now be made from steel, a stronger material than iron. Manufacturers now specialized in precision bearings, which offered tighter fits and better control. Ball bearings remain a fundamental engineering control in the friction reduction toolbelt. You'll find them in gear boxes, bicycles, and even wind turbines.

In an ideal world, engineering controls, like incorporating ball bearings into a system, would be sufficient to reduce friction. In the real world, however, not all moving parts can be connected via low-friction ball bearings. The forces acting on the bearing might be too high, or the machine might experience vibrations whose noise is amplified by ball bearings, or the cost of the ball bearing assembly might be prohibitive. We need other solutions to achieve low friction. As much as we'd like to just throw water on a system and make parts more slippery, that isn't always practical. Some of the most challenging friction problems share qualities we discussed in Chapter 2: dry contact between solids.

Just as anoles increase friction by creating more contact zones, tribologists can minimize friction by reducing contact between solids. Sometimes, they manipulate the mechanisms of friction, such as adhesion, by altering the attraction between surfaces or making it easier to shear them apart. Sometimes, they reduce the plastic deformation of asperities, either by altering the material properties of the contacts or the operating conditions. At other times, such as when launching a satellite into orbit around Earth, it's not possible to change the operating conditions. Fortunately, solid lubricants keep things moving.

Solid Lubricants

Referring to lubricants as solid may seem strange. Lubrication almost always refers to liquids, like water or oil. However, solid materials can also be lubricious. Our first introduction to this happens as children, when we start scribbling with a crayon. Crayons date back to the sixteenth century, although the eponymous crayon brand, Crayola, wasn't founded until the turn of the twentieth century. The significance of friction in crayons is hidden in the name of the company itself. The Crayola company's founders, Edwin Binney and C. Harold Smith, chose the name "Crayola" on the advice of Alice Stead Binney, Edwin's wife and a schoolteacher, who had suggested they create a cheaper alternative to imported crayons. Mrs. Binney joined two French words, the word *craie,* which means chalk pencil, and *oléagineux,* which means oily, to coin the word Crayola: oily chalk. Considering that the earliest crayons were oiled charcoal, later pigmented to create different colors, oily chalk is certainly an apt name.

Designing crayons involves carefully balancing friction. The friction from sliding a crayon along a piece of paper needs to be high enough to soften the wax and transfer the color to the paper. At the same time, it needs to be low enough that there's no stick slip, which would cause the color to be transferred in a staccato manner. Additionally, friction between the layers of wax needs to be low enough to enable transfer at all. This is possible because of the weak intramolecular forces between the layers, which allow the layers of wax to slide past each other and transfer to the paper.

Crayola crayons, like most crayons today, are made from paraffin wax, but with all things involving friction, the formula is continually optimized. Nearly a hundred years after the launch of Crayola, patents are still being filed for new formulations intended to improve the transfer of color while also minimizing the annoying flaking that can happen with crayons. These formulations contain careful blends of fatty acids, hard waxes, and soft waxes.[6]

That we haven't run out of crayon formulations, even after hundreds of years, highlights the huge variety of natural and synthetic waxes available. You are likely familiar with plant- and animal-based waxes, such as soy and beeswax, which are used in candles and lip balms, and paraffin wax, used in candles and cosmetic products. Derived from petroleum, oil, or shale, the latter has come under scrutiny in recent years as consumers wish to be less reliant on fossil fuels. Synthetic waxes, created from the chemical synthesis of polymers, include polyethylene and polypropylene wax. These blends are used to manipulate properties like hardness and melting point. However, the polymers are derived from petroleum feedstock, so recent work on biopolymer waxes has focused on reducing petroleum reliance.

Depending on the type of wax used and its application, friction can be reduced by adjusting the moisture content of the wax, whether through wicking or by separating two surfaces with a layer of wax. Winter sport enthusiasts wax their skis and snowboards to minimize the friction on snow, and cyclists wax their chains to keep pedaling smoothly. Of course, on the opposite end of the spectrum, surfers use waxes to increase the friction between their feet and the board so they don't go sliding off. Waxes also help reduce the friction between personal protective equipment (PPE), such as face masks, and our skin. A thin layer of wax can go a long way toward making PPE more comfortable and reducing skin irritation by reducing static friction.[7] When selecting such waxes, tribologists consider the thermal properties, such as the melting point, hardness, and stiffness of the wax material, as well as its environmental compatibility, including its toxicity.

There are three traditional solid lubricants: graphite, molybdenum disulfide, and polytetrafluoroethylene. The definition of a solid lubricant can be contentious. Some tribologists define solid lubricants as those with low friction, but this is rather ambiguous since friction is a system property. Others define them as materials that produce a low-friction film between contacts. Yet others may point to the microscopic structure of the material itself. The pencil contains one of the most common solid lubricants, one that satisfies all three of these definitions.

As a kid, I was constantly bewildered by the fact that what we call pencil lead is not lead at all. Until almost 1800, graphite was mistakenly identified as black lead. The Romans used lead styluses to write. But in the Middle Ages, Europeans realized that graphite was darker and more legible, as well as easier to transfer. They named it *plumbago,* a word derived from the Latin word for lead, *plumbum.* Later, when they realized that it wasn't lead at all, they renamed it graphite to distinguish it from lead. *Graphein* in Greek means "to write."

The pencil was invented by a French chemist named Nicolas-Jacques Conté. He mixed graphite, water, and clay to form a material soft enough and with low-enough friction to be transferred to paper, and then fired it in a kiln around a wooden barrel to solidify and harden the mixture. One of the reasons graphite works so well is that it's soft, despite having the same chemical composition as diamond, one of the hardest known materials.

Both graphite and diamond are composed of carbon. How can pure carbon produce both graphite and diamonds? The answer lies in the arrangement of each material's atoms. Graphite has a hexagonal crystal structure. Six carbon atoms form a hexagonal ring, and the rings form layers, held together by a combination of weak covalent bonds and van der Waals forces. The weak forces holding the layers together enable them to slide readily over each other. The layers are self-lubricating, meaning an external lubricant supply isn't needed to maintain low-friction performance.

Diamond, on the other hand, has a face-centered cubic crystal structure, so each carbon atom shares a covalent bond with four other carbon atoms. The carbon atoms in graphite share only three covalent bonds with neighboring carbon atoms. The stronger bonds in diamond make it much stronger and harder. They also prevent layers of atoms from easily sliding over each other. As a result, diamond isn't self-lubricating.[8]

The second of the traditional three solid lubricants, molybdenum disulfide (MoS_2), was often mistaken for lead and graphite. As you may have guessed, MoS_2 is also a shiny, silvery material that looks very similar to pencil lead. In fact, the name molybdenum derives from the ancient Greek word for lead, *molybdos.*[9] Molybdenum has been found on the moon, Mars, and meteorites, but it doesn't occur naturally

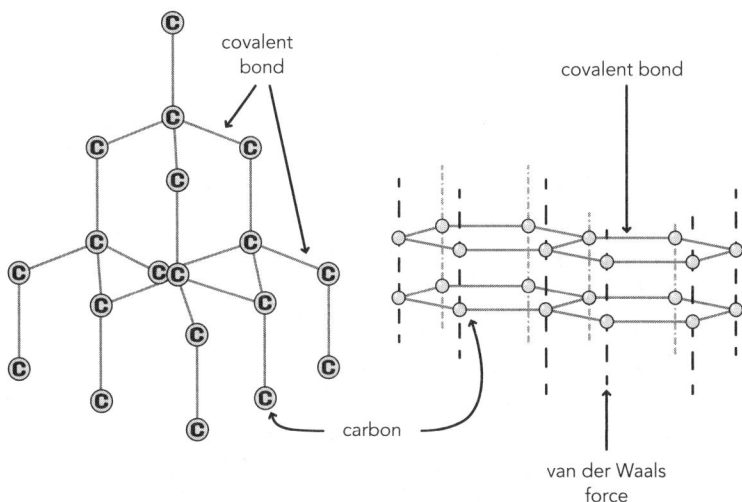

covalent bond

covalent bond

carbon

van der Waals force

The chemical structures of diamond and graphite

on Earth.[10] Some researchers believe the existence of molybdenum on Mars may be proof that life started on our Milky Way neighbor. Molybdate, a highly oxidized molybdenum, is considered essential for life, preventing organic substances from turning to tar and protecting RNA from corroding in the presence of water. It requires more oxygen than is available in our atmosphere, so researchers have suggested that billions of years ago, it traveled to Earth from Mars via meteorites.[11] Others have suggested that it helped oxygenate our atmosphere, enabling life to evolve.[12] Whether we are technically Martian or not, we know that molybdenum is present in the oldest microorganisms: archaea and bacteria.

Because molybdenum does not exist as a pure element on Earth, it was difficult to identify initially. It would take a trio of Swedish chemists in the late 1700s to isolate it. While molybdenum doesn't occur naturally, molybdenum disulfide (MoS_2) does.[13] Like graphite, MoS_2 gets its lubricious nature from its similar hexagonal chemical structure. In this case, the layers are more complex, as there's more than just carbon. A hexagonal layer of molybdenum atoms covalently bonds to a layer of sulfur atoms above and below it. These layers of S–Mo–S then stack on top of each other, creating two sulfur layers held together by weak van der Waals forces. With this arrangement, it doesn't take much shear strength to separate the S–Mo–S layers and transfer one layer to another surface, providing self-lubrication, just like with graphite. Because the layers of MoS_2 can slide over each other so easily, the friction between them is lower than the friction between MoS_2 and another surface.

Given the similarities between graphite and MoS_2, you might assume they can be used interchangeably. They can't. They have different properties. The former is an organic material that can be produced from living organisms; the latter is an inorganic one. A major difference between the two materials is that graphite needs to have its layers terminated by hydrogen to be lubricious. This is another way of saying that the edges of the crystal structure need hydrogen to slide easily. The edges have high energy and are considered dangling bonds, meaning they will readily bind to other materials. This can lead to adhesion and high friction. If the dangling bonds are exposed to water, however, the hydrogen in water will readily attach to those high-energy sites, preventing adhesion. If you rub graphite against another surface, transferring it to that surface, its edges are continually exposed. Throughout civilization, this wouldn't have been an issue for anyone using graphite as a lubricant. Water, with its two hydrogen atoms, is abundant in the atmosphere, so when operating in air, graphite is lubricious. However, in vacuum environments, where there is no water, graphite is not the material you want.

MoS_2, on the other hand, thrives in a vacuum. Molybdenum readily bonds with oxygen, or oxidizes, and becomes nonlubricious molybdenum trioxide (MoO_3). The less oxygen available to MoS_2, the better. That's why it's the go-to material for lowering friction in a vacuum or in low-oxygen environments. This oxidation issue also sets temperature limits on MoS_2. Molybdenum on its own can tolerate significantly higher temperatures of up to 2700 °C before it will melt. However, at elevated temperatures, MoS_2 in air will interact with oxygen, decomposing into MoO_3 and SO_2. This limits its service temperature to below 400 °C. To benefit from the temperature capability of MoS_2, you need, once again, to be in a vacuum. By contrast, graphite can operate as a solid lubricant up to 450 °C in an oxidated atmosphere, although its friction increases around the boiling temperature of water. The environment and operating conditions are incredibly important in the selection of graphite versus MoS_2.

Solid lubrication is not what most people immediately think of for reducing friction, but when the operating environment is too hostile for a traditional lubricant like water or oil, a material that won't burn or freeze is needed. It's no coincidence that significant progress in the development of solid lubricant systems occurred at the same time as a critical innovation in transportation: the turbojet engine. These new jets were much faster and could cover greater distances due to an internal combustion chamber that offered more thrust than a traditional propeller. Combustion chambers reach over 500 °C, but traditional transmission fluid ignites at 195 °C. Solid lubricants like graphite and MoS_2 were the solution.

In 1938 scientists discovered what may now be the most controversial solid lubricant ever used: polytetrafluoroethylene, shortened to PTFE. Often referred to by the trademarked brand name it was introduced under, Teflon, it was not developed with the intention of lowering friction. At the time of its discovery, Roy Plunkett, a researcher with the chemical company DuPont, was working on refrigerants. Plunkett was an Ohioan who rose out of poverty to earn a PhD in chemistry before embarking on a career as a researcher at Dupont. At the time of his work, refrigerants were toxic and flammable. Plunkett wanted to develop an option that was preferably neither. To do so, he started tinkering with tetrafluoroethylene (TFE) gas, which is somewhat ironic given that current TFE data sheets warn of the highly explosive nature of TFE gas when it interacts with oxygen. It's also listed as a possible carcinogen. Of course, these are things we learn with time and data.

Plunkett hypothesized that reacting TFE with hydrochloric acid would yield a suitable alternative to the common refrigerant Freon.

He stored the gas in cold canisters, cold enough to preserve dry ice. Then, when he was ready to start the experiment, which involved mixing chlorine with the TFE gas, he opened a valve to allow the TFE gas to flow from its cylinder into the experimental canister. Except the TFE gas didn't flow. His experiment had failed.

When Plunkett and his assistant, Jack Rebok, tipped the TFE cylinder upside down, a white powder sprinkled out. As they scraped the sides of the cylinder, they realized that they had created a waxy substance. Despite the fact that scientists had thought it impossible, Plunkett had polymerized the TFE, forming long chains of a polymer by combining smaller molecules. The TFE gas had reacted with iron in the cylinder once the pressure was high enough to catalyze the reaction, and PTFE was born.

PTFE has a backbone made up of thousands of carbon atoms with strongly bonded fluorine atoms attached to it. In fact, fluorine forms the strongest bond with carbon. Fluorine is highly electronegative, meaning it holds its electrons tightly. This, combined with its small size, means it pulls carbon's electrons close enough to itself that the fluorine and carbon electrons overlap, creating a short, strong bond that is inert to other chemicals. Plunkett found that PTFE had useful properties: in addition to high temperature resistance, corrosion resistance, and low surface energy, it was slippery. Moreover, when he tried to destroy the wax with heat and other chemicals, it didn't change. This substance was unlike anything they had seen before.[14] In 1941 DuPont patented PTFE.

Today, Teflon is most associated with nonstick pans. But its first application was in the Manhattan Project, perhaps portending the future of PTFE itself. The Manhattan Project, which produced the first atomic bomb, used some nasty chemicals, such as the highly reactive and corrosive uranium hexafluoride, to separate isotopes of uranium for fuel. Scientists needed a material that could be used to coat valves and that wouldn't be destroyed by or react with chemicals, causing an explosion. PTFE was conveniently chemically inert. The new material caught the eye of General Leslie Groves, the US Army general serving as director of the Manhattan Project. Teflon, and thus PTFE, became a classified material, and commercial applications would have to wait. But when the war ended and this new polymer was declassified, DuPont wasted no time in trademarking the Teflon name and putting it to commercial use. The first major application was the nonstick pan. Because PTFE is inert, the material could prevent anything from sticking to it, from super-secret classified substances to eggs. PTFE was used in fabrics for resistance to stains and water. And the National Aeronautics and Space Administration (NASA) famously used it in the Apollo program as a coating on the astronauts' spacesuits to protect them from harmful radiation and to keep abrasive moon dust from destroying the suits. PTFE was a game changer, and it seemed to be everywhere.

A seemingly miraculous material at the time, PTFE has now become a contentious one. You've likely heard of PFAs, also called forever chemicals because they don't break down in the environment or in our bodies. PFA stands for per- and poly-fluoroalkyl substances. In other words, it's a fluorinated polymer, meaning it's primarily composed of carbon-fluorine bonds. The acid that was used to make PTFE, a PFA called PFOA or perfluorooctanoic acid, is a fluorinated material found in the environment. While not a manufacturer of PFAs, DuPont relied heavily on PFOA in the production of Teflon. For decades, PFOA was dumped into the Ohio River Valley, exposing communities in West Virginia and southern Ohio to toxins that have been connected to serious health issues, including cancer. A highly publicized class-action lawsuit resulted, inspiring a Hollywood telling of the story. Both raised awareness of the dangers of making PTFE and the long-lasting environmental impact of PFAs. Ultimately, DuPont exited the Teflon business and, along with two other companies, Chemours and Corteva, settled the case for over $1 billion in June 2023. A PFA producer, 3M, faced other lawsuits from a number of cities and states for contaminating drinking water. In the weeks following the DuPont case announcement, 3M settled with the city of Stuart, Florida, for over $10 billion. The concerns around PFAs arose nearly eighty years after

Plunkett discovered PTFE. Until then, PTFE was a highly regarded material with intrinsically low friction.

Engineering with Solid Lubricants

Although solid lubricants took off in the 1940s, they were hardly a new discovery. Applying them to machinery dates to the ancient Egyptians, who lubricated wheels with traditional substances such as olive oil and tallow, but also solid ones, including lime powder and calcium soaps. Humans, however, didn't invent solid lubrication. Nature beat us to it. Beetles, for instance, have a protein in their leg joints that acts as a solid lubricant when dry, keeping friction at the joint as low as it would be if PTFE were there instead. Earth's plates have solid lubrication, too. For hundreds of millions of years, glacier-scrubbed sediment has coated Earth's crust, controlling the grinding and churning of plate tectonics responsible for earthquakes and tsunamis.

In the 1940s, as engineers pushed technology beyond the confines of Earth's gravity with the development of the turbojet engine, they needed to figure out not only what solid lubricant to use but how to apply it to parts. The parts encountered environments in which liquid lubricants would evaporate, causing equipment stalls and failures. The solution started with one patent, a 1944 filing detailing how to bond graphite to metals to create low-friction coatings in engine pistons. This was a critical step forward because, as it turns out, a material that has intrinsically low friction doesn't like to stick to things.

Designing solid lubricant systems that will stick to materials like a frying pan is a science in itself. There's the lubricant selection, method of delivery, and design of the resulting system. Each step in the process is as important as the others. PTFE is a good example of some of the ways tribologists can play with solid lubricants. PTFE can start as Plunkett found it, in powder form. The powder can be compressed, heated, and molded into solid parts like blocks and rings. It can also be made into a fiber form, as Wilbert Gore and his son Robert discovered when they heated and stretched a rod of PTFE. And really stretched it. They managed to stretch the PTFE well past 800 percent of its original size, creating expanded PTFE or ePTFE. You may know this by its trademarked name GORE-TEX. Jackets and clothing made from GORE-TEX are prized for their breathability while still being waterproof, thanks to ePTFE fibers.

These different forms of the same material come in handy when optimizing a system for whatever it is you want out of it. When using a solid lubricant, like PTFE, you want low friction. Odds are, however, that you also need other properties and performance out of the material. Maybe you need the material to be a certain strength or hardness. Often, in tribology, low wear is desired along with low friction. After all, you want that low-friction material to last as long as possible. For all its other impressive properties, PTFE has terrible wear performance, which limits its usefulness as a solid lubricant on its own. A successful solid lubricant needs to transfer to another surface, creating a film or coating on that surface that reduces friction in the system. If the transferred layer, called a transfer film, can be rubbed off easily, low-friction performance is jeopardized. The challenge with PTFE is that it doesn't stick to materials very well because it is chemically inert. In fact, it usually comes right off. To make matters worse, it flakes off in large chunks, allowing for rapid wear, which means that PTFE doesn't last long. Tribologists design PTFE-containing materials that balance the desired properties of low friction, low wear, and temperature and chemical resistance.

Blends of materials such as these are known as composites. The bulk, or majority material, is known as the matrix, and the other materials included are fillers. Composites aren't limited to just two materials. There are plenty of composites out there with multiple fillers. For example, you'll find polymers that contain carbon fiber for strength and PTFE for lubrication. The possibilities are endless and are often driven by economics. If a lower-cost matrix can be used and its properties boosted with fillers, the final product will save the customer money, making it a winner. Cost of manufacture and economics are always in the equation in material design, though sometimes, depending on the performance requirements, a lower-cost matrix isn't possible.

While not the cheapest material option, low-wear PTFE composites can be made by adding PTFE powder as a filler to another material, like a high-performance engineering polymer. Polyetheretherketone (PEEK) is a polymer with excellent thermal and mechanical properties, but it can have a high coefficient of friction in moving systems. Blending PEEK powder with PTFE can create a low-wear, low-friction material. A robust transfer film forms when enough PTFE is present in both the film and PEEK composite material to provide the low-friction benefits of PTFE. Meanwhile, the PEEK polymer chains are able to prevent the high wear of PTFE. This type of low-wear, low-friction composite can also be made by adding expanded PTFE fibers to a material, like PEEK. Alternatively, other materials such as carbon or glass fiber, which have high strength and good wear resistance, can be used as a filler to strengthen and reduce wear in PTFE-containing products.

A small amount of PTFE can drastically reduce the friction in systems. Likewise, a small amount of filler can improve the wear performance of PTFE by orders of magnitude. In my PhD lab, those of us who blended together materials for low wear and low friction were called the lab "bakers." Careful, precise measuring, mixing, and, well, baking of materials results in optimum performance of the composites.

The Science of Coatings

As desirable as it is to have an entire part made from a composite material, sometimes that's not possible. Sometimes, it's not even the best solution. For example, polymer composites are temperature limited because kinetic energy causes their bonds to vibrate, leading to breakage. This results in the polymer melting or degrading. If the application requires high-temperature performance beyond a polymer, a ceramic or metal is necessary. Maybe an engineer wants to launch something into space and needs it to be temperature resistant as well as lightweight. They'll want to choose the lightest metal or ceramic possible because weight savings are mission critical. Unfortunately, ceramics can create more friction than desired. The solution is to apply a coating to one of the interacting surfaces, which is exactly what French engineer Marc Gregoire did when he invented the first nonstick cookware.

Humans have used coatings for tens of thousands of years. The red and black pigments used to adorn Paleolithic caves with depictions of anthropomorphic figures and animals contained oxides of iron and manganese. These compounds were mixed with fats or plant matter that acted as binders, holding the pigments together and helping them stick to cave walls. The ancient Egyptians took painting to new heights, developing the first synthetic color—"Egyptian blue"—by mixing sand containing lime and silica with copper and natron and then heating the mixture in a furnace.[15] They also worked with transparent, tinted coatings known as varnishes. These they made from linseed oil, which hardens as it interacts with oxygen in the air.

The variety of painting techniques developed in the ancient world reflects a sophisticated mastery of coating application. Egyptian painting was done in *secco,* or dry overpainting. Painters would mix pigments with binders and apply them to the dried work surface. The binder would then harden the pigment onto the surface to set the painting. In Greece, Bronze Age Aegean artists worked with the *fresco* technique, painting on a wet surface. Archaeologists have spent decades working out the technical details of these paintings, employing many of the analytical tools tribologists use to understand coatings, including microscopy, scanning electron microscopy, and X-ray techniques that can identify chemical components. Early in the Bronze Age, mud was the primary plaster medium, serving a functional purpose. It was sticky enough to adhere to walls, breathable, and available in abundance. Later, plaster was mixed with lime, which created a smooth, white surface. The ancient Greeks discovered that adding pigment to a plaster surface while it was still wet works brilliantly to set the colors. As it dries, the calcium hydroxide in lime migrates to the surface, where it reacts with carbon dioxide in the air. This forms calcium carbonate crystals on the surface, locking the painting into place. Greek artists may not have understood the finer details of this carbonization process, but they knew to take advantage of it.[16]

We often discuss the evolution of technology in terms of scientific progress, but innovation in the behemoth paint and coating industry was driven first by art. Coatings protected materials such as wood, which would otherwise succumb to elemental damage from the environment. The need to work with, preserve, and beautify wood led to advances in varnishes and polishes. These include shellac coatings, which were first developed in the thirteenth century. Hundreds of years later, we still use shellac coatings, made from shellac resin, in varnishes. Resin is produced by the female lac bug, a scale insect found in the forests of India and Thailand. The lac bug clings to twigs and branches and consumes tree sap. After digesting the sap, it secretes a wax through its pores. The wax then reacts with oxygen in the air, hardening and forming a shell to protect the eggs the lac bug is laying. This hardened shell can be scraped off a tree to produce shellac resin. Heating up shellac resin causes it to melt, creating a clear, glossy coating that can protect artwork. Pigments and metal powders can be added to create glossy, colorful coatings. Well before shellac, artisans used lacquer, another natural varnish produced from the sap of lacquer trees. Japanese and Chinese artists began lacquering their pots in 7000 BC to create brilliant, glossy finishes. In the seventeenth century, European carpenters adopted the practice, and by the 1800s, spurred by the Industrial Revolution, the first paint and varnish factories had opened.

Today, protection against corrosion and temperature resistance are the primary drivers of new coatings technology. Reducing friction is another. Often called antifriction coatings, coatings designed to lower friction appear in a variety of applications, spanning many industries. The reasons for using a coating can vary and often overlap with the reasons for using a solid lubricant: the need for clean operation without oil and grease contamination, the problems of high-temperature environments where oils and greases combust or degrade, and the demands of corrosive conditions. Graphite, MoS_2, and PTFE are all used in antifriction coatings but blended with other materials to achieve the desired set of properties. Antifriction coatings, like the pigments used to paint cave walls, often need binders. Ideally, the binder can help meet other performance needs, but it doesn't have to—its primary job is to hold the ingredients together and ensure the coating adheres to the material. Selecting a binder material is a critical step in developing coatings and must take into account the properties of the coating material itself, the material the coating must be applied to, and how the coating will be applied.

The molecular structure and chemical composition of the binder dictates how it interacts with other materials. We've seen this already. For example, PTFE would be a terrible binder because the fluorine atoms prevent it from bonding to other materials. The lime in plaster, however, can react with carbon dioxide in the air at room temperature, making it an ideal binder. When you purchase a coating, it's common to see a curing time and temperature listed on the container. Lime needs time to react and cure, but it can do so at room temperature. Other binders need heat to bind to the substrate material. Tribologists must also consider how the binder will affect the dispersion and mixing of the coating itself. A binder that adheres easily to another material isn't helpful if it also causes the solid lubricants to clump together and form an uneven coating.

Another consideration when selecting a binder is the coating technique to be used. Painting is the easiest, and oldest, method of applying a coating, but it's not necessarily the best. Spraying and dipping aren't much better. It's hard to control the thickness with these methods, and coatings often need to be uniform and tightly controlled in order to meet design requirements. Parts are always designed to what is called a *tolerance,* meaning the wiggle room in the dimensions. If you're coating a gear in an assembly and the tolerance is exceeded, you could jam up the entire system. In space and aerospace applications, tolerances are extremely tight, so you can imagine how thin coatings are. Furthermore, every milligram of mass counts, so thin coatings are a necessity to prevent extra weight. It's not unheard of for antifriction coatings to be only a few atoms thick.

One way to create such a thin film is through vapor deposition. Vapor, the gas phase of a material that usually exists as a solid or a liquid, can be deposited onto a surface a few layers of atoms at a time. Vapor deposition techniques are split into two categories: physical vapor deposition (PVD) and chemical vapor deposition (CVD). In PVD, the material source is physically applied to the surface, whereas CVD relies on chemical reactions to apply the coating. You might think that a device that applies a material in atomically thin layers would itself be small. It's not. I first saw a PVD machine in graduate school, when I was touring tribology labs in the United Kingdom. The machine was a massive silver box, about five meters long and one and a half meters wide, with a handful of knobs sticking out of it. This enormous machine was required to coat a sample half a meter long and half a meter wide.

Looking behind the doors of a deposition coater reveals the complex processes that necessitate such a large machine. Solid material has to be vaporized, transported to the material being coated, and then condensed onto that material, all under vacuum conditions. PVD techniques are multistep processes that evaporate solid material into a spray of small particles to create a coating. The machine does this by applying heat or by bombarding the solid coating material with a high-energy electron beam. The kinetic energy of the electrons is converted to thermal energy that is high enough to sublimate the particles from solid to gas, resulting in a vapor cloud. This cloud then needs to be transported to the coating chamber, where the substrate is located. Typically, this is done through pressure differences in the machine. During vaporization, the particles move from a higher to a lower pressure chamber where the substrate is mounted. The vaporized atoms are then deposited, one atomic layer at a time, onto the substrate, where they condense, creating a thin coating. All of this has to occur in a vacuum in order for the particles to travel to their intended target.

The drawback of PVD is that it involves physical interactions. The surfaces being coated need to be uniform and free of contaminants. Any variations in the surface can result in a nonuniform coating. CVD, on the other hand, is more forgiving of irregular surfaces since it involves chemical reactions. Also applied in a vacuum, CVD coatings are grown atomic layer by atomic layer through chemical reactions with the surface or with a chemical reagent. As with PVD, the source coating material is solid. CVD sometimes requires a precursor chemical to initiate the reaction, so the temperatures in the chamber must be tightly controlled. Otherwise, you won't end up with the reaction you wanted. CVD is widely used by the semiconductor industry to create silicon dioxide coatings, which are used in the ubiquitous chips and wafers found in our phones, computers, and a variety of electronic devices. One problem with CVD is that while the tightly controlled reactions of CVD tend to produce more uniform coatings than PVD, the process involves chemicals and processes that may not always be environmentally friendly.

CVD has been around since the dawn of civilization, when torches used by early cave dwellers would deposit soot on the walls.[17] It was window glass, however, that made CVD critical in manufacturing. During the 1970s oil and gas supplies were restricted, the result of an embargo on the United States by Arab nations. Roy Gordon, a chemist and professor at Harvard University, believed that nations shouldn't be so reliant on importing fuel. He wanted to reduce energy waste by developing more thermally efficient window glass. Despite a lack of interest from the US government, Gordon pressed on, developing and testing numerous coatings. A tin oxide coating, both transparent and conductive, was the winner. His naysayers said there was no way to apply the coating to large sheets of glass. CVD, he found, was the solution. It just needed to be scaled up for production quantity and speed.

It took ten years and a dedicated team of researchers to successfully achieve this. The process takes hot sheets of glass as they exit a furnace and floats them over a bath of molten tin. A coating machine above the glass releases gases, which, due to the high temperature, react with the molten tin to create tin oxide. This clear material then hardens onto the glass to form the coating. Today, this coated glass is produced by nearly all glass manufacturers, and the savings it provides on heating bills in just one year can cover the wholesale cost of the glass. CVD is also used in a variety of energy applications, including solar panels, bringing us one step closer to Gordon's dream of achieving an energy-independent nation.[18]

The usefulness of CVD coatings isn't limited to glass or electrical conductivity. One technology that has benefited from CVD is MEMS. MEMS, which stands for microelectromechanical systems, was touted as one of the most exciting technologies of the twenty-first century and received much attention and fanfare. As the name suggests, MEMS include both mechanical and electrical components, ranging from springs and levers to electrical devices such as resistors, capacitors, and inductors. MEMS are typically under a thousand micrometers in size, although some are on the order of millimeters. The quest for smaller and smaller systems means there now exists an even smaller generation, called nanoelectromechanical systems (NEMS). MEMS were first developed for commercial use in the 1960s, but in the 1990s they experienced enormous growth and market penetration, thanks to manufacturing advances in the electronics industry, including the production of microchips found in computers and phones.[19]

These tiny devices are now fixtures in our lives. Your phone contains multiple MEMS, from the microphone to the accelerometers that automatically orient your screen and track your physical activity. The first big commercial application of MEMS was in airbags, where they are still used to sense rapid deceleration and a force threshold. When patients receive intravenous fluids, MEMS can be used as a wearable sensor to monitor blood pressure. In fact, sensors are a major application of MEMS. Usually, friction is less of a concern since such devices are used in static conditions. But MEMS can also move. They're used in actuation, to convert electrical energy to physical movement. You'll find microactuators in robotic surgery grippers, camera focusing mechanisms, displays, and inkjet printers.

Just because the movements of MEMS are small doesn't mean that friction isn't an issue. On the contrary, friction, or stiction, as people in the MEMS industry often call MEMS friction, can still be a hindrance. Unfortunately, the smaller the parts, the trickier it can be to manipulate friction. Coulomb's first law of friction, $F = \mu N$, applies to all macroscopic systems. But as we saw in Chapter 2, it doesn't always apply to all atomic-, nano-, and microlevel systems. The best way to understand friction at such small scales is to measure it directly. Scanning probes such as AFM can be useful at these incredibly small scales. These microscopes made an appearance in our discussion of contact mechanics, but to really appreciate them, and how they can be used to measure atomic forces, it's worth exploring atomic force microscopy in more detail.

The first question to ask is how to measure a force. A common way to do this is to measure displacements caused by the applied force. This requires relating force to displacement. One approach is to use Hooke's law, which relates the stiffness of a spring and change in length of the spring to the applied force. We often associate this with classic coiled springs, but it also applies to components like cantilevers, the beams or spans of material fixed to one end of the AFM probe. In the case of AFM, the cantilever stiffness is known. As the microscope tip scans along the surface of a sample, the lateral displacement of the tip is measured. The amount of displacement is related to the stiffness of the tip; a stiffer tip will compress less than a more flexible one. Since the displacement of the tip is incredibly small, it is measured with light. A photodiode, a semiconductor device that generates an electric current from light, is used to reflect light from the beam on the tip. The resulting electrical resistance can then be used to determine the distance the probe has moved and provide measurements sensitive enough to measure surface forces at the atomic scale.

With its high sensitivity, AFM has unlocked a new world of friction measurements, enabling quantification at the nanoscale level.[20] Diamond-like carbon (DLC) coatings, which can be just nanometers thick, provide one such example. AFM experiments have shown that at low loads friction is governed by interfacial forces like van der Waals forces, but at higher loads, it becomes more dependent on load and looks like "$F = \mu N$" again, adhering to Coulomb's first law. With the insights AFM experiments have provided, tribologists can design coatings composed of numerous materials to create the recipe of properties they're after. Then, using CVD, they can apply coatings to the complex, small geometry that is MEMS. At one time, MEMS switches were unreliable, sticking due to friction. This hampered their commercial success. Now, thanks to tribological work such as antifriction coatings, not only do MEMS switches account for nearly half the MEMS market, but antifriction-coated MEMS can be in contact with other moving parts in hard drives, microphones, printers, and myriad other applications without causing damage to the devices.[21]

DLC coatings have facilitated the digital revolution by solving tribological challenges with disk drives. DLC coatings offer excellent wear resistance in the form of thin films. The amount of storage contained on hard drives is exponentially higher when compared to a decade ago. Since the first version of the hard drive was introduced by IBM in 1956, storage has increased annually anywhere from 30 to 100 percent.[22] The average hard drive in 2018 had 10^7 times more storage capacity than the original 1956 drive. Inside a hard drive is a magnetic disk platter coated in a thin film of magnetic material. The platter consists of sectors that can be independently magnetized. Each sequential change in the direction of the magnetic field represents a data bit. As the disk platter spins, a thin magnetic arm, called a read-and-write head, scans each side of the disk. This arm converts the platter's magnetic field to an electric current when reading the disk or converts an electric current into a magnetic field when writing to the disk. Speeds are high, on the order of ten meters per second, and distances between parts are on the order of nanometers. DLC or thin carbon coatings are used to protect and lubricate the surfaces. These distances have become so small that there is often only a single atomic layer deposited on the surfaces. The ability to protect surfaces and maintain low friction while reducing the distances between parts has enabled higher storage volume, faster speeds, and more reliable drives.

DLC coatings were a game changer for the electronics industry. But more recently, silicon-based coatings have been gaining traction. These coatings combine excellent conductivity, temperature performance, and mechanical and tribological properties. Tribologists have also been exploring thin polymer-based coatings. These coatings are possible thanks to CVD, during which the molecule vapors are polymerized into a film on the substrate.[23] The diversity of processing techniques, coupled with the material options, allow for countless combinations of coating formulations to achieve performance results in a variety of applications. The antifriction coating market is pretty darn healthy, with a predicted growth rate of nearly 8 percent between 2020 and 2028. The expected market in 2028 is over $1.5 million. Even with all the recent developments, MoS_2, PTFE, and graphite still dominate the market.[24] Bear in mind, however, that often, coatings will include more than one of these lubricants. It's common to find MoS_2 and graphite together since they each perform well under different conditions—MoS_2 under vacuum conditions and graphite in a high-oxygen environment. A tribologist might as well combine them to try to get the best of all worlds, especially for parts exposed to a wide range of temperatures and environments, such as in space.

Surface Engineering

Coatings fall under the broader category of surface engineering, a suite of techniques that engineers use to manipulate the properties and performance of moving parts. The key point is that the engineering occurs only at the surface. The bulk of the material below the surfaces layer remains unchanged. One way to do this is by applying a coating. However, sometimes, as we just saw, the material surface itself must be modified just to apply the coating. Even the ancient artists did this, using mud and lime plaster in Aegean frescoes. Coatings are just one type of surface engineering. There are other ways to modify surfaces that don't involve applying a coating.

Instead of creating a coating, it's possible to alter the chemical composition or physical properties of a material. Physical surface modifications occur any time we ski or snowboard, a discovery made almost by happenstance by Philip Bowden, introduced in Chapter 2. Bowden realized that high contact pressure in the real contact area zones would lead to high temperatures in those zones during motion. This frictional heating could lead to localized melting. Water melting is essential for skiing. In a brilliant application of the adage "do what you love and never work a day," Bowden combined his favorite hobby—skiing—with his research. During a ski trip to the Jungfrau in Switzerland, when a blizzard left Bowden and his companions snowbound, he channeled his boredom into pondering the friction between his skis and the snow. Bowden found it puzzling that he could ski down the mountain even though the snow was below –5 °F. With solid-on-solid contact, it seemed improbable that friction could be so low. But once he got his skis moving, they slid down the snow with ease.

Osborne Reynolds, yet another famous scientist known for his work outside the realm of tribology, had surmised that this was due to melting from the pressure of skis on the snow. However, this explanation was debunked when it was determined that even the weight of an elephant couldn't create the melt needed to reach the low friction skiers experience. The answer, Bowden eventually found, was that frictional heating caused by the motion of the skis against the snow was high enough to melt a very thin layer of the snow, providing liquid lubrication. This small surface modification is a physical one that has a large impact on friction without changing the chemical composition of the material itself, in this case, water. In Chapter 4, we'll explore the various ways liquid lubrication, such as melted ice, can reduce friction.

Other modifications do change the chemical composition of the surface layer. You might have heard of anodized aluminum, found in cookware and appliances, or carburized steel, used in gears and bearings. Both are examples of this type of surface modification. Guy Dunstan Bengough and John McArthur Stuart introduced anodization at an industrial scale by developing and patenting a technique to anodize aluminum with chromic acid (H_2CrO_4). The aluminum part is added to a bath of chromic acid, and a voltage is applied. This voltage generates the electrical current needed to start the chemical reaction for the process. The acid releases oxygen, which interacts with the aluminum part to create a hard, oxidized protective layer on the surface.[25] Most commonly used with aluminum to prevent decay and corrosion, anodizing can also be used on other metals like titanium or magnesium. The thin oxide layer prevents the otherwise very reactive metal underneath from reacting with its environment. Without anodization, aluminum would be useless in many everyday applications that we take for granted, such as in kitchen appliances and building exteriors. In addition to protecting the aluminum, the oxidized layer binds well with dyes, providing a more adhesive surface for applying paint or creating a glossy finish. Anodizing is increasingly important as vehicles and aircraft move toward lighter-weight designs, replacing steel with aluminum to reduce energy consumption.

The downside of friction-modifying materials is that wear is inevitable. If a coating wears away, the friction of the system will change abruptly, sometimes resulting in catastrophic failure. If the modified surface wears too much, modifications like an oxide layer are removed. That's why, in general, engineers prefer to make an entire part out of a material with low-friction properties than apply a coating to the surface. But even uncoated materials can experience frictional changes from wear, as we saw with the PTFE transfer film. As a result, tribologists must balance wear and friction. Wear is a subject in its own right, with entire textbooks devoted to it, so the details won't be explored in this book. But it should be apparent that optimizing a system to achieve steady friction of a desired amount requires engineers to understand the wear mechanisms of that system.

Skiing and other snow sports enthusiasts rely heavily on all the tricks tribologists play to control friction and wear: engineering design, material selection, and surface engineering. Tribologists are continually optimizing winter sports equipment. The design of skis, snowmobile track liners, and snowboards has evolved from planks of wood to cambered designs to prevent increased resistance at the front tips. The materials have also evolved considerably, from exclusively wood parts to wood centers with steel edges to polymer composites. Today, Olympians use skis made of polyethylene bases with carbon fiber and fiberglass fillers to reinforce and adjust their stiffness and strength. Polyethylene is hydrophobic, meaning that it repels water, so there is minimal interaction with the melting snow, further reducing friction.

Skis illustrate various surface engineering techniques. Grooves of different depths, placed at strategic locations along the skis, can control the contact with snow and the flow of melted water to yield lower friction. Waxes can further facilitate gliding and minimize friction, although in some cases, waxes are used for the opposite effect: to improve gripping by increasing friction. Because friction is influenced by snow conditions and the speed and weight of the skier, there are seemingly endless combinations of designs and lubricants available. Elite skiers have dozens of skis designed to optimize friction in different weather conditions.

For snow and motor enthusiasts, snowmobiles are also designed with friction in mind. Some users prefer to invest in snowmobile slides designed for combined low friction and low wear. These somewhat expensive liners have wear-resistant inserts to reduce maintenance intervals and are coated in solid lubricants to achieve low friction. I used to work on these liners and enjoyed trawling through forums discussing their merits; many a lively debate has erupted over these versus more traditional ones that have a lower up-front cost but wear out faster.[26]

Traditional Lubrication

One day, while taking a break from biographing friction, I was watching a cooking program. The chef instructed the viewer to throw cornmeal on the counter before handling the dough to "reduce friction." Even cornmeal is a solid lubricant, and chefs must be tribologists! Of course, with cooking, despite cornmeal and nonstick pans, most of us also grease the pan to mitigate friction. Oil and grease are the common fix for most of our friction challenges. After all, when your bike chain becomes stuck or your door is squeaking, you don't grind up your pencil lead. No, you probably go for the can of WD-40 or specialist bicycle grease.

Tribologists refer to liquid lubricants like WD-40 as traditional lubricants. They are our default for lowering friction, and their use dates back thousands of years to pottery wheels that spun easily when wet with water from the clay. In fact, whenever I tell people I'm a tribologist and explain what tribologists do, they immediately mention WD-

40. There's a famous engineering flowchart with the question "Does It Move?" at the top and yes or no options underneath. If it is moving but shouldn't, the solution is duct tape. If the answer is no, it isn't moving, but it needs to move, the solution is WD-40. And, yes, WD-40 has come into play in my own tribology experiments. But as ubiquitous as that lubricating spray is today, it wasn't adopted immediately when it first appeared. The name itself gives away its backstory: Water Displacement 40th. As in, the fortieth attempt to create a water dispersant.

In 1953, a trio of scientists from the Rocket Chemical Company were working on a solution to prevent corrosion and rusting in the aerospace industry. Their plan was to create something to disperse water, which would cause undesirable corrosion. In a proper homage to the saying "If at first you don't succeed, try, try again," and again—thirty-eight more times—they finally invented WD-40. The name was inspired by annotation in the chemists' notebook and was eventually shortened to WD-40.

The product was highly successful, helping to protect the Atlas Missile, the first intercontinental ballistic missile, from rusting. Amusingly, it was such a good product that some of the engineers took it home for their own use. The Rocket Chemical Company decided to make an aerosol can for home use. Eventually, in 1969, the company renamed itself WD-40, and the rest is history. Pretty much every household now has that distinct blue and yellow can sitting on a shelf somewhere. Its uses have also expanded. WD-40 has been used in a variety of unexpected applications, from removing gum from clothes to removing a burglar stuck in an HVAC duct. WD-40 works to prevent corrosion, which minimizes friction, but even without the rusting issue, it acts as a lubricant to reduce resistance in moving parts.

Highly confidential, the formula for WD-40 is known to only a handful of people. We can surmise that it's composed of a base oil, but beyond that, the magic is hidden behind the blue and yellow can. Many liquid lubricants, however, follow a generic recipe of a base oil and what are known as additives, used to improve or provide the desired properties of the material. The base oil, which is often either a mineral or synthetic material of hydrogen and carbon but can also be a biological material, separates the solid surfaces to reduce friction. Additives, typically a very small proportion of the overall formulation, are used for a variety of purposes, including to influence the friction and flow of the oil and to prevent rusting. If you own a car, you've probably noticed the various options available to you for oil changes. These products advertise the latest, greatest synthetic oil with proprietary additive packages guaranteed to make your car practically magical.

Grease is another traditional semi-solid lubricant that is often added to lubrication formulas. It acts as a thickener, altering flow properties. In this way, traditional lubricants are like solid lubrication. Just as a pinch of a low-friction solid lubricant can make a big difference, so too can a small number of additives in oils and greases. In some cases, the additives may even *be* a solid lubricant. The fundamental operation and physics of solid lubrication versus traditional lubrication is, however, quite different.

Thus far we've comfortably spent time in the world of solid mechanics. But to manipulate friction with liquids, we have to explore the physics of fluids. It is through an understanding of fluid behavior that we can see why skis have microgrooves on the surface to prevent water build-up, reducing friction. It's how we progressed from trains on tracks to maglev trains, letting air become our medium instead of metal-on-metal. And, perhaps counterintuitively, it's how we know that there is such a thing as too much lubricant in a system. This brings us to the world of fluid mechanics and lubrication theory, the topic of Chapter 4.

4 *Going Against the Flow*

To AVOID THE OPPRESSIVE heat and humidity of Floridian summers, one of my favorite pastimes as a kid was watching Wimbledon. I'd spend my mornings sprawled on the floor in my pajamas with cookies in hand and a cat by my side, enthralled by the deft strokes and line-hugging shots of the players. I love everything about tennis—the cerebral, tactical maneuvers, the mechanics behind executing the perfect drop shot, and especially the science of designing the equipment. Sliding on the court is an important aspect of the game, and perfecting it can make or break a match. Friction, of course, governs how a player's slide will be executed. Wimbledon offers a perfect storm to make sliding unpredictable: grass.

Throughout the tournament, the state of the grass evolves as it is worn down from use. The fickle British summertime weather also affects the condition of the courts. Moisture on the court grass, even a small amount, can change the contact of the player's foot with the ground. A small amount of moisture can be enough to induce unintended slipping, ranging from minor skids to a full wipeout on the court. Controlled sliding is part of the game of tennis, but losing traction completely is not. These accidental slips on the grass are concerning for players, even if they result in some media-worthy photos. In some years, extensive slipping becomes the main story, as more and more players tumble, injure themselves, or even withdraw from the event. Tennis shoes are designed to provide enough friction to keep the player from falling while permitting some degree of sliding. The catch is that the shoes must optimize friction performance in the presence of potential lubrication from moist grass. The friction in question isn't simply between two solid surfaces; it also involves a fluid—water.

So far, we've focused on friction between solid surfaces, but the time has come to tackle that second type of friction: fluid friction. It perhaps goes without saying that the physics of fluids is significantly different than that of solids. Whereas solids resist being sheared, fluids continuously deform under external forces. Fluid friction, also called lubricated friction, often refers to interacting systems with liquids such as water, oils, and greases present. But fluid friction in the broadest sense includes liquids *and* gases. (And plasma, but we won't go there in this book!)

As fluids deform under external stresses, the atoms and molecules shift around. This produces internal friction in the fluid, called viscosity. Viscosity is a fluid's resistance to flow, or put another way, its resistance to deformation. The higher a fluid's viscosity, the more it will resist flow. Honey is often used as the example of viscosity; it is much more viscous than water, explaining why it's harder to squeeze it out of the bottle. There is also a force acting opposite the bulk fluid's motion, such as when the ocean flows against the hull of your boat or the air against your car. This resistant force is known as drag. For your sanity and, frankly, mine, we'll separate liquids and gases as we dig into the world of drag, viscosity, and all things fluid friction.

Liquids

Fluid friction falls under the broader branch of physics known as *fluid mechanics,* a fascinating and complex discipline concerned with the behaviors and properties of fluids when forces act on them. Until now, we've been working with classical mechanics of rigid, discrete particles, but fluid mechanics is a subdiscipline of continuum mechanics, which treats materials as continuous rather than composed of discrete particles. Properties such as velocity, pressure, and temperature vary continuously in fluids. What we observe is the average of many particles. Rather than trying to figure out what each molecule or atom is doing, we choose a representative volume of the fluid and apply continuous functions of space and time to describe properties and behavior of the material.

Isaac Newton was the first to mathematically describe viscosity, in the second book of his three-volume work *Principia.* Whereas the first volume focused on motion without resistance, the second concerned motion within media of resistance, like a liquid. "Resistance" is the key word signaling that we're dealing with friction. Newton proposed that fluids require a force to be sheared because of internal friction. His law of viscosity states that the shear stress between fluid layers is proportional to their velocity gradient. Today, we call a fluid "Newtonian" if the relationship between viscosity and shear stress is linear, that is, if it is constant and independent of the rate of shear stress over time. Water is the most familiar of Newtonian fluids, but alcohols and other fluids are also Newtonian. Non-Newtonian fluids, whose viscosities change with the amount of shear stress applied over time, include paint and toothpaste.[1] Newton may have been the first to document viscosity, but two other men are independently credited with laying the foundations of our current understanding of it: Jean Léonard Marie Poiseuille and Gotthilf Hagen. They both measured the flow of fluids through pipes of various sizes, developing the first instruments to measure viscosity. That two men with vastly different backgrounds—one studied medicine and the other was a civil engineer—arrived at the same conclusions highlights how influential this property of fluids is.

Poiseuille's name is synonymous with fluid mechanics, but the figure behind the name is largely enigmatic. We know that at the age of eighteen, he entered the prestigious École polytechnique in Paris, an institution with a reputation for scientific excellence. However, the school was swept up in political turmoil shortly after Poiseuille enrolled. In 1816, when a handful of students protested in support of the republic, the school was temporarily shut down. The following year, King Louis XVIII reopened the school, but Poiseuille opted not to return. What Poiseuille did next is unclear, but we know he studied medicine and presented his dissertation to the faculty of medicine at the Sorbonne in 1828. He earned his doctorate on aortic forces, and his thesis stated that he was a former student of the École polytechnique.[2] Details of where his work was conducted, and who sponsored it, are lost to history.

Interested in hemodynamics, the flow of blood through vessels, Poiseuille wanted to know how artery size and pressure affect flow.

To answer this question, he measured how fast water flowed through glass tubes of various diameters.[3] Poiseuille controlled the pressure by filling to different heights a water reservoir that supplied the tubes. Using a hand pump, he pumped air into the reservoir and measured the resulting pressure. After reaching the desired pressure, he would open a valve to the tube, allowing water to flow through it. The tubes themselves were small, but the size of the reservoir made the overall apparatus a couple of meters tall. Measuring the change in the height of the water, Poiseuille was able to calculate the flow of the water through the selected capillary tube diameter. After obtaining replicable results and perfecting his device, he repeated the experiments with other liquids at various temperatures.[4] From these experiments, Poiseuille determined that fluid flow through a tube increases in proportion to applied pressure and to the diameter of the tube raised to the fourth power.

Poiseuille had developed a method to measure viscosity in a controlled manner. The device he designed was an early viscometer, an instrument still used today to measure viscosity. Whether he appreciated it or not, he had created controlled flow, also called *laminar flow*. Such flow is found in healthy blood vessels. In contrast, *turbulent flow* is caused by chaotic changes in pressure or velocity and produces visible eddies and swirls. Today, the results of these experiments have become known as Poiseuille's law, and the metric unit for viscosity is called the poise.[5] Poiseuille's law relates the volumetric flow rate (Q) to the radius and length of the tube, pressure across the tube, and viscosity of the fluid for laminar flow. It's written as $Q = (\pi r^4 P)/(8\eta L)$. If you know the viscosity of a fluid η, you can predict the behavior of its flow through capillaries of any length L, radius r, and pressure P. Poiseuille's law explains why high blood pressure occurs when our arteries become constricted and is critical for determining the flow rate of intravenous fluids.

Around the same time Poiseuille was studying blood, Gotthilf Hagen, a German civil engineer, independently derived the same relationship between pressure, flow rate, viscosity, and tube diameter. Hagen specialized in hydraulics and water management. He first worked as a port inspector and later managed the hydraulic engineering of many of Germany's rivers. Flow and pressure changes across different-sized pipes are important in hydraulic engineering, directly affecting how efficient a system is and how much energy is required to pump water through the system. To better understand these effects, Hagen created his own viscometer, regulating the pressure of water at the entrance of a tube, just as Poiseuille had done. In this case, however, the tubes were brass instead of glass and an order of magnitude larger. Hagen published his work in 1839, whereas Poiseuille would publish his findings shortly after, in 1840. Today, this work is known as the Poiseuille-Hagen law in recognition of the ambiguity around who published this result first.[6]

Hagen and Poiseuille both invented what is known as capillary viscometry to make their measurements, flowing liquid through a tube and varying the pressure and diameter of the tube. Today, capillary viscometers offer a cost-effective way to measure viscosity. However, the most common type of viscometer is the rotational viscometer, first developed in 1890. This instrument has two surfaces that rotate relative to each other, shearing the fluid contained between them.[7] When the instrument allows for measurements across a wide range of conditions and measures more than just viscosity, the instrument is known as a rheometer. In addition to viscosity, rheometers measure the flow and deformation of materials, including properties like elasticity and yield. They can be used to study properties like *thixotropy,* that is, how the viscosity of a material decreases with increasing applied stress, or how different formulations affect the manufacture, performance, and even our sensory perception of a product. The discipline that involves studying the flow and deformation behavior of materials is known as *rheology.*

The properties of fluids measured by rheometry subtly influence your everyday life. Our bodies have plenty of examples of rheology in action. Muscles, for example, are thixotropic, which means they become less viscous when under more stress. Synovial fluid, which lubricates cartilage where it comes together to form joints, and connective tissue relax as we move, enabling them to flow more easily.

The more we move, the more the fibers are able to stretch and relax. As they rest, they gradually constrict and stiffen again. This is why it's so important to warm up before exercising and take care to cool down afterward.[8]

Characterizing viscosity and flow properties under different conditions is also important for ensuring the shelf life and stability of products. Over time, a complex liquid like a shampoo may experience structural changes in its materials as the molecular chains settle to a lower energy state. As the structure of the liquid changes, with networks of molecules either connecting to each other or breaking apart, the internal friction of the fluid changes, affecting its flow. This happens to many products, hence that pervasive "shake well before use" label. Understanding how viscosity changes is especially important for developing medications and administering vaccines, for which the last dose of medicine needs to be identical to the first dose. With viscosity measurements providing a map of the fluid under different conditions, tribologists can provide guidance, such as shaking products before use or recommending storage temperatures to avoid separation.

Rheology measurements guide *how* liquids are made, ensuring that the ingredients are evenly dispersed. They also inform how materials are applied during manufacturing. It's easy to mess up a coating by applying it at the wrong speed, which depends on viscosity. Rheometers have a lot of ground to cover. Like tribometers, they have to be run at conditions that match the application, whether it's chewing, mixing, or coating. That's why you'll find a variety of testing configurations and instrument sensitivities. In fact, some rheometers are sensitive enough to detect the equivalent of dust on the end of a meter stick.

Rheological measurements are especially useful for understanding time-dependent flow behavior, such as the aforementioned thixotropy. Thixotropy is the property of materials that enables them to be poured or spread more easily after being shaken or squeezed. Thixotropic behavior is a big part of developing gels, which we find in everything from cosmetics to food. In fact, the viscosity behavior of gels in food directly influences your culinary experiences. You'd be amazed at how many food companies have rheometers on hand to characterize their products and ensure your experience is as palatable as possible! Ketchup is another fluid designed for thixotropic behavior. As you apply more force to the bottle, the ketchup flows more easily. You can see the same effect with toothpaste, lotions, and sauces. These fluids need shape, but not so much shape that they can't be spread around. If their viscosity didn't decrease with force but set when left alone, well, have fun with that mayonnaise. You might find that it won't spread, or it might just run off your bread. Printing inks are another great example of the importance of viscosity and thixotropic behavior, something Gutenberg figured out in the 1400s. His mechanical type printing press would never have succeeded without his ingenuity in materials science. At the time, water-based inks were used for printing, which meant they ran, making the type illegible. Gutenberg developed oil-based inks, more like paint, that were thicker and had higher viscosity. He mixed linseed oil with walnut oil, lampblack, pitch, and other oils until he landed on the optimal combination of viscosity and color. Today, the viscosity of inks must be lower on printer rollers than on the printed surface. Once printing is complete, the ink needs to set, returning to its high-viscosity state until it has dried completely.

We've only scratched the surface of the vast field of fluid mechanics. We've discussed viscosity and how we measure it, but in order to quantify any of the fluid's properties with a viscometer or rheometer, we need to understand the physics behind fluid properties. As mentioned earlier, there are different types of fluid flow. The type of flow impacts the formation of a lubricating fluid film. And this is where we revisit a name mentioned earlier: Osborne Reynolds.

Reynolds and Fluid Flow

Osborne Reynolds was born in Belfast in 1842 and attended Cambridge University, where he studied math and mechanics before entering the workplace as a civil engineer. His work in industry as an engineer was short-lived, however. At the age of twenty-six, he was appointed a professor at the University of Manchester. A brilliant man, his quirks as a professor are well documented: forgetting about lectures and running in late, challenging classes with lessons that were difficult to follow, deriving theorems he thought might be wrong in the textbook, only to decide that, yes, they were in fact correct. But Reynolds was also known to be friendly, humorous, and personable. Respected and popular among his students, he embodied a combination of eccentric and kindly professor stereotypes. What he wasn't keen on, however, were academic pretense and self-promotion.[9]

As a researcher, Reynolds liked to tackle practical problems. In that sense, the civil engineer in him was never far away. Growing up, he apprenticed with a shipbuilder, sparking his lifelong interest in fluid mechanics.[10] He had a knack for taking an amazingly complex topic and distilling it down to mechanical principles easy enough for all to grasp. Even today, the ability to unpack complex science into digestible concepts is one of the most important skills tribologists develop. This gift, along with his penchant for tackling real-world problems, led Reynolds to become an eminent figure in fluid mechanics, and especially lubrication. His contributions were so significant that he has both an equation and a number named after him.

Curious about both the practical and philosophical side of science, Reynolds wanted to expand the work of the fluid mechanists before him. He set his sights on Newton's law of viscosity. In 1883, he asked what would happen if he flowed water through a glass pipe and injected water with dye into the mix. A valve at the end of one pipe allowed him to control the flow velocity, much like when you use a garden hose today. He noticed that at low velocities, the dyed water maintained its shape as it traveled through the pipe. When the velocity was high, the dyed water would break apart. What Reynolds observed was turbulent flow. He then repeated the experiments at different temperatures. This allowed him to investigate the impact of viscosity as the molecular behavior of the water molecules changed. He observed that the ability of the fluid to break apart wasn't just dependent on speed but also viscosity. This makes sense since viscosity is the fluid's internal friction or ability to withstand shearing.

Reynolds noted that a combination of viscosity and speed would cause the transition from laminar to turbulent flow. The Reynolds number measures fluid flow: it is a ratio of two fluid properties, resistance to motion and internal friction, or viscosity. In general, when the Reynolds number is above 3,000, flow is turbulent, and when it falls below 2,300, flow is laminar.[11] For values in between, flow is considered turbulent as it transitions from laminar to fully turbulent. The Reynolds number has important implications, affecting how a fluid transports heat. A higher Reynolds number means the fluid is moving faster, and the heat transfer rate will be higher than a fluid with a lower Reynolds number. One of the consequences of laminar versus turbulent flow is how fluids mix. When flow is laminar, the layers of the fluid don't mix because the flow is streamlined. If you're trying to mix fluids, then you want turbulent flow. But if you want to prevent mixing, then you want a system that maintains laminar flow. Ventilation systems used in hospitals and labs employ laminar flow. The disorderly nature of turbulent flow would kick up dust and particulates in the ducts, leading to potential bacterial growth.

Lubrication

Viscosity influences a fluid's ability to perform as a lubricant. This isn't surprising given Reynolds's observations of viscosity's effect on fluid flow. In 1885, Reynolds published a nearly eighty-page treatise on fluid lubrication. Lubrication, he explained, is the process by which viscous fluids such as oils "diminish friction and wear between solid surfaces." But at the time, as Reynolds noted, it was an area that had been given scant theoretical treatment. People used lubricants but didn't understand why they worked. It was an area ripe for experimentation.[12]

Reynolds wanted to model a phenomenon that an English engineer named Beauchamp Tower had observed. Tower had run friction experiments on a journal bearing, a sleeve that supports a rotating shaft, to simulate railway axles. He showed that journal bearings work by being separated entirely by a fluid film. In making this observation, Tower had stumbled upon *hydrodynamic lubrication,* an ideal situation in which two solid surfaces are separated by a thin film and friction is minimized. Reynolds wanted to figure out how fluids could flow into the space between a journal bearing and shaft, a small gap under pressure. He worked with all sorts of configurations: stationary parallel plates with fluid flowing between them, one plate moving, both moving, one angled. He found that the fluid was "dragged" into the contact by the relative motion of the surfaces and that the resulting thin lubricating film would carry the load applied to the bearing. The summation of all this work gave rise to what is now known as the Reynolds equation, which describes the flow pressure distribution of thin films across a bearing. The equation allows us to determine the thickness of a fluid film and how the design of pressure affects the ability of a fluid to lubricate a surface.[13]

The Reynolds equation is a beastly looking thing with multiple terms to account for the multiple directions in which fluids can flow (x, y, and z). It relates the viscosity, velocity, and dimensions of a bearing to the hydrodynamic pressure and, ultimately, the performance of the lubricant. It explains how the lubricant can enter the space between two surfaces and provide the thin film necessary to reduce friction. It's also the basis of lubrication theory.

Lubrication Regimes

It's no coincidence that efforts to understand the lubrication of a basic machine component, such as a bearing, occurred during the Industrial Revolution. At a time of soaring production, understanding lubrication became a necessity. Economies were tied to the output of factories and mills, to the heavy machinery that cranked away within their walls. In the mid-1870s, an engineer named Robert Henry Thurston connected friction to the inefficiency of mill equipment. A successful factory relied on good lubrication. Thurston's work resulted in a crucial observation: the speed and force of two surfaces pushing against each other affect the efficiency of a lubricant.

Unlike many of the other scientists and engineers discussed so far, Thurston was not a Brit. Born in the United States, he studied civil engineering at Brown University. He went on to become one of the most renowned mechanical engineering professors in the country and the first president of the American Society of Mechanical Engineers (ASME). He was a member of the Navy Engineering Corps during the Civil War, then spent time working at the Naval Academy before being appointed chair of mechanical engineering at the Stevens Institute of Technology in New Jersey. Like other pioneers of friction, including Coulomb and Reynolds, Thurston eschewed traditional academic research, which he regarded as too far removed from practical problems. He was more interested in the industrial applications of his research, applying math and science to real-world problems. At Stevens, he overhauled how mechanical engineering was taught. He pioneered the transition to the modern curriculum taught to mechanical engineers like me—a mix of coursework and lab work. Theories taught in the classroom, Thurston believed, should be tested in the lab. In fact, he founded the first mechanical engineering lab in the United States, where he developed machines to investigate practical research problems. There, friction and lubrication took center stage.[14]

To Thurston, friction was the enemy, the foe that engineers needed to defeat. Perhaps, given his influence over the curriculum, this is why friction continues to be taught with the same bias today. When Thurston examined cotton and flax mills, he realized that about 50 percent of available power was being used to overcome friction. The resulting economic losses were significant. Thurston lectured around the world on the issue, published well-received papers and addresses, and, crucially, ran tests to help understand the impact of friction on machinery.

He also built machines, such as the Thurston Oil Tester, also known as Thurston's machine. Designed in the 1870s, it was one of the first lubricant testing machines.[15] It consisted of a horizontal shaft running through two bearings. At the end of this shaft was an overhanging bearing, which suspended a pendulum. Inside the pendulum was a spring that maintained the desired pressure on the bearing. This bearing was used to test the lubricating ability of oils. As the horizontal shaft rotated at a set speed, the pendulum would swing to a different height, depending on the friction in the oiled bearing. The friction between the bearing and shaft provided the torque needed to swing the pendulum. If the bearing was, in theory, frictionless, the shaft attached to the crank would spin freely and not translate any motion to the pendulum. High friction between the bearing and shaft would essentially lock the two parts together such that turning the crank attached to the shaft also turned the shaft of the pendulum until the pendulum was horizontal. Therefore, the smaller the swing, the lower the friction in the bearing.[16] With this instrument, Thurston could test the friction between the shaft and various oils under different loads and speeds. He created a table of the conditions and the results, providing a handbook on lubrication for engineers. A century before the establishment of tribology as a discipline, Thurston raised the importance of friction, linked it to economics, and presented a practical approach to figuring out what to expect from machinery under different conditions.

However, Thurston didn't plot his data. It's amazing how much a graph can elucidate the previously unknown. Thurston's tables suggested that friction reached a minimum value and then, at certain speeds and pressures, crept back up. He had, unknowingly, discovered the transition from partially lubricated contacts to fully separated, hydrodynamic lubrication.

In the 1880s, German engineer Adolf Martens noticed that friction undergoes transitions under different lubrication conditions. By plotting his data in graphs, he generated curves that provided instant visual feedback on the performance of a lubricated system as conditions changed.[17] These curves filled in missing data in Thurston's work, showing how to minimize friction by modifying the velocity, pressure, and viscosity of a lubricant. Still used by tribologists to select lubricants, these curves demonstrate distinct regions of behavior, known as lubrication regimes: friction starts high, drops off, then increases

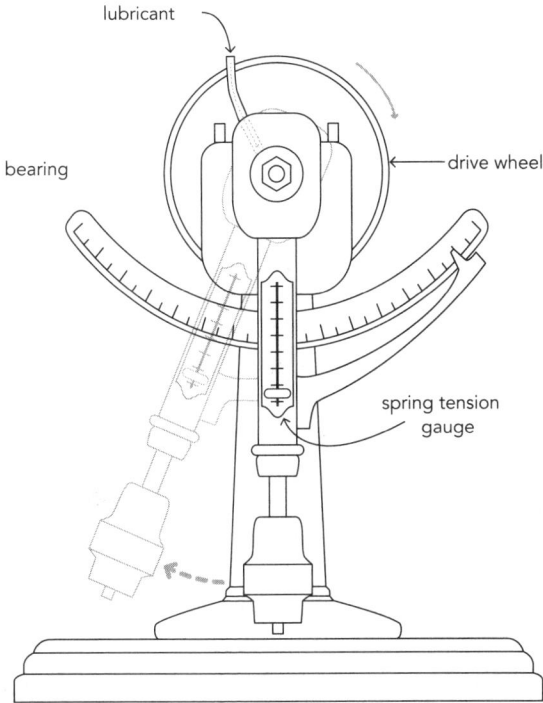

Thurston's lubrication oil tester

slightly as any or all three parameters change. Unfortunately for Martens, this important tool is known as the Stribeck curve. Richard Stribeck was a German engineer who was studying lubricated surfaces of journal bearings. Like Martens, Stribeck varied pressure and velocity and recorded the resulting friction.[18] The difference was that Stribeck published his results in a more prestigious journal. As a graduate student, I may not have fully appreciated why journal impact factors are so important, but I certainly do now! Make sure you choose your publisher wisely, or your Martens curve may be named for someone else.

About a decade after Stribeck plotted his curves, Ludwig Gumbel incorporated viscosity into the plot. Viscosity, as we've seen, is one of the most important properties of a liquid. The tricky thing about viscosity is how temperamental it is. Viscosity changes under different conditions, the most notable being temperature. For example, it's much easier to spread warm butter on bread than cold butter. That's because there's more resistance to shear. At higher temperatures, liquid molecules move around more quickly, reducing the intermolecular attractions between them. This is sometimes problematic—especially in a machine—where frictional heating of nearby parts can alter a lubricant's viscosity, changing its lubrication regime and resulting in higher friction and wear. Cars used to need winterized fluids because fluids would become more viscous in cold temperatures. Now, thanks to a better understanding of lubrication, oils and fluids are engineered to withstand a range of temperatures. Engineers carefully characterize the viscosity of liquids, varying temperature, pressure, and shear rates to ensure optimal performance. There are even standards for classifying lubricants to make this work easier for the user.[19]

Gases, on the other hand, become more viscous with temperature. This has been especially important for applications such as petroleum engineering, in which the viscosity can change the efficiency of production by resisting the transport of oil or natural gas. While gases and liquids both follow the Reynolds equation, different states of matter yield different characteristics that affect lubrication performance.

The pressure and shear rate of fluids also alter the viscosity of fluids. For low pressures and low shear rates, viscosity remains nearly constant. Under high pressure, liquids tend to become more viscous. However, as the shear rate increases, liquid viscosity decreases. That's because as the molecules stretch, they become less resistant to shearing. You can permanently change the viscosity of a fluid by stretching it until the molecular chains break.[20]

By incorporating viscosity into the mix, Gumbel's version of the Stribeck curve shows how load, velocity, and viscosity interact, affecting the coefficient of friction. If the load and viscosity are held constant, we can see how friction will respond to velocity. As mentioned in

Chapter 2, fluids break the third law of friction and exhibit velocity dependence. This can be seen when the coefficient of friction is plotted against a lubrication parameter called the Hersey number.

The Hersey number relates the viscosity of a fluid to the velocity and applied load of the system—parameters that, thanks to Reynolds's work, we know influence fluid film thickness. To appreciate the Hersey number, it's helpful to see its equation written out:

$$H = \frac{\eta V}{P}$$

The Hersey number increases with increasing viscosity, η, and velocity, V, and decreases with increasing pressure, P. It's considered a measure of fluid film thickness, with a thicker film forming at higher Hersey numbers.

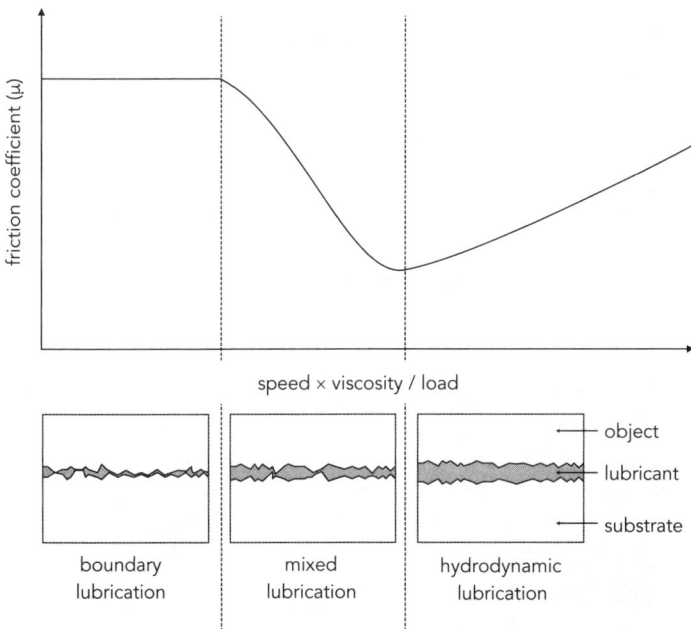

Classic Stribeck curve

In the Stribeck curve, the velocities cover orders of magnitude. For this reason, the curve is often plotted on a logarithmic scale. In experiments where the velocity is the only parameter varied, the curve is simply plotted against the velocity, as shown above. Friction is highest on the left-hand side of the graph. This lubrication regime is known as *boundary lubrication* and yields high friction because it's mostly solid-on-solid contact. Imagine it's drizzling outside, and a couple of raindrops land on the ground beneath you. There's not much water present, but some water is lubricating the sole of your shoe. As the rain becomes heavier and more lubricant becomes present between the solid contacts, in this case, the treads of your shoe and the ground, the lubrication regime shifts to mixed lubrication, meaning both boundary and hydrodynamic lubrication can occur, depending on the conditions outside, the speed of your gait, and the pressure you apply to the ground. Exactly how these conditions influence the lubrication regime and friction depends on the exact materials involved, but typically with higher pressure and higher velocity, a system transitions from boundary to mixed to hydrodynamic lubrication. While boundary lubrication isn't usually desirable because of the higher friction, I think mixed lubrication is the worst regime because of the range of friction behavior you can get as you transition between the other two regimes. As the system wears, some of the surfaces may suddenly not be in contact, and the friction drops precipitously. It's just a little too unpredictable and unstable for my liking. Eventually, minimum friction occurs when a very thin fluid layer separates all of the solid contacts. While this isn't what we'd want while walking in the rain, it is the sweet spot engineers aim for in hydrodynamic theory.

One of the biggest success stories involving the Stribeck curve can be found under the hood of a car. Using the curve, tribologists can optimize the efficiency of equipment based on parameters they can control. The history of automotive transmission fluids emphasizes the power of this. These fluids consist of base oils and performance additives that stabilize them over a range of temperatures. One of the purposes of these fluids is to reduce friction by separating the moving engine components with a layer of low-friction lubricant. The catch is that fluids have drag, stemming from their viscosity. Engineers and scientists must find the correct balance, modifying formulas to create a fluid that yields a thin lubricating layer, which the driver then experiences through increased fuel efficiency. It's estimated that lower viscosity transmission fluids can increase fuel efficiency by 2.5 to 5 percent.[21] This may sound modest, but when you consider how it scales across millions of cars, it adds up quickly! Of course, while lower viscosity lubricants might offer the possibility of energy reduction, they also create design challenges. Thin films increase the likelihood of contact between moving parts, which in turn increases friction. While the optimum regime is easy to identify on a Stribeck curve, achieving it isn't always so simple.

One problem with the Stribeck curve is that hydrodynamic theory is based on conformal contact, which means the surfaces in contact conform to each other. Often conformal contacts are made of one convex surface paired with a concave surface. Picture a shaft passing through a bearing, whose sleeve shape is designed to fit around the shaft, or our hip joint, where the spherical ball at the top of the femur fits into the socket at the pelvis. Now picture two balls in contact or two gears contacting at their teeth—these don't conform to each other at all. The contact is nonconformal, and so the contact area is much smaller. A smaller contact area means higher contact pressure than for conformal contacts. Many common machine parts are nonconformal. So how do we know when nonconformal contacts are fully separated by a fluid film?

Elastohydrodynamic lubrication (EHL) theory plugs this gaping hole. The development of EHL theory is a tribology tale that includes espionage and intrigue. It begins in the 1930s, when the Soviet scientist Alexander Mohrenstein-Ertel recognized that lubricant films between nonconforming gear contacts do not follow the Stribeck curve. Under classical hydrodynamic theory, the high pressures of nonconforming contacts should result in boundary lubrication and high friction. However, high pressure does two important things not accounted for by lubrication theory. First, it causes metal contacts to deform elastically, as Hertz showed. Second, it causes the lubricant viscosity to increase, thickening the film and separating the surfaces. EHL theory takes these two factors into account. Just as Martens's curve was named after someone else, Ertel's method was also credited to another engineer—his professor, A. N. Grubin.

This was not because Grubin was power crazy or Ertel had failed to publish his work. After World War II, Ertel was sent to East Germany to lead a friction and lubrication lab. Accounts of what happened next are, well, interesting. Ertel defected to West Germany, assuming a new identity and working as a scientist in Schlewecke.[22] However, the Soviets spread other stories. According to one account of events, Ertel traveled to West Germany to gather scientific information and vanished, possibly killing himself. Whatever the true story, his work was not widely published. His former boss, A. N. Grubin, realized Ertel's work was too important to just gather dust, so he published it himself. EHL theory was long credited to Grubin. It's only recently that the behind-the-scenes drama has come to light, and Ertel's efforts have been acknowledged.

Today, tribologists use a version of the Stribeck curve that includes the EHL regime. You'll notice that the x-axis is either the dimensionless parameter of the Hersey number, the velocity, or a new variable, λ. This variable, called the lambda ratio, is the ratio of the lubricant thickness to the surface roughness. When the lubricant film thickness is less than the height of the solid asperities, the asperities will touch. This is what occurs in boundary lubrication, where friction is dominated by dry, solid contact. As the lambda ratio increases, the lubrication regime moves from the boundary to the hydrodynamic regime. When λ is greater than 3, a lubricant is in hydrodynamic territory; less than 1, it's in boundary territory. Everything else is in the mixed lubrication regime. If the contacts are nonconformal, EHL occurs when $3 < \lambda < 10$. Greenwood and Johnson, whom we met in Chapter 2, were important contributors to the development of this ratio.[23] It makes perfect sense that the scientists who devised a way to quantify and model surfaces would figure out how to apply surface modeling to lubricated surfaces.

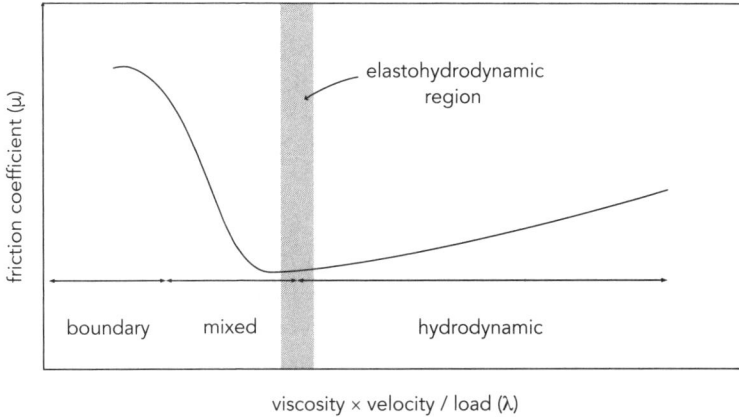

viscosity × velocity / load (λ)

Stribeck curve with elastohydrodynamic lubrication regime

The Art of Lubrication Measurement

The obvious question when using the lambda ratio is this: How do you know what the film thickness is? One option is to measure it. The problem is that you can't just measure film thickness by holding up a measuring tape between a bearing and shaft while these parts are spinning around in a machine. Even if you could, you'd be assuming the thickness is uniform around the entire shaft, which isn't likely.

The interference of light waves has long influenced tribology research. Color is one of the most powerful indicators of light interference and can be used to measure film thickness. This is called interferometry. In 1933, Katherine Blodgett, a chemist at General Electric (GE), working with the future Nobel Prize–winning chemist Irving Langmuir, developed a method for measuring film thickness based on its color. Blodgett was the first woman hired to work at GE's research labs and would also become the first woman to receive a PhD in physics from Cambridge University. She was developing a method to deposit monolayers of barium stearate and calcium stearate, soaps that can also be used as heat stabilizers and lubricants, onto a glass plate. While doing so, she observed that the color of the glass plate changed with increasing layers. As white light travels through a medium of different layers, the proximity of each layer can cause the light to interfere. This interference will change the wavelength of the light, making it longer or shorter. Blodgett realized she had invented a highly sensitive ruler.[24] The most sensitive devices at the time were capable of measuring a few thousandths of an inch; Blodgett's monolayers were orders of magnitude smaller. Blodgett had created the first interferometer, a color gauge based on the lubricant's sheen that indicates the thickness of the film.[25] With this color gauge, thin coatings of known and uniform thickness could now be made. The impact of her work extends beyond lubrication. Barium stearate–coated glass is one of the least reflective glasses. Blodgett found that with forty-four monolayers, 99 percent of light would pass through.[26] For this reason, barium stearate–coated glass was used in the filming of *Gone with the Wind*, a movie considered innovative for its clear picture.

EHL films are so thin that they can be measured with wavelengths of visible light. In fact, light waves are a perfect tool for measuring EHL film thickness. If we shine white light on the contact region, the reflected wavelength shifts through various colors based on the film thickness—just as Blodgett's ruler had. The overall result is a lubrication map showing the film thickness during operation. An oil film color map looks similar to a heat map, except the various colors represent film thickness instead of temperature.[27] Such maps were a breakthrough in lubrication theory when developed in the 1950s, enabling researchers to verify Ertel's EHL equations and predictions of EHL film thickness. Over the decades, tribologists have optimized the method. Today, it's possible to measure EHL film thickness down to the one-nanometer range.[28] The method is also used to measure boundary lubrication, where small amounts of lubricant are present in otherwise dry contact regions. Tribometers are fitted with cameras focused on the contact between the sliding surfaces so they can image the films during operation.

Tribologists have coupled interferometry with other techniques, gaining additional insight into what happens to lubrication during use. Laser-induced fluorescence is one such technique that can be used to visualize lubrication. It is the practice of adding a dye to a material that then fluoresces when exposed to certain wavelengths of light. The energy of the laser excites the fluorescent material, which then emits light particles, called photons, to a detector. This is a powerful technique for visualizing flow, and it's been used to study lubrication of prosthetics, windshield wipers, and shaving razors, among others. Yes, shaving. Most modern safety razors have a lubricating strip that helps reduce friction during shaving, preventing cuts.

Suzanna Whitehouse, a researcher at Imperial College London, has studied how the lubrication strip changes with use and how it interacts with water—important factors in determining the materials for and design of the lubrication strip.[29] I can recall wondering years ago why the lubricant strip was there since it just washed away after the first use. Thanks to laser-induced fluorescence, we can see how different lubricant strip formulations perform by mimicking the system on a tribometer. By marking the lubricant with a fluorescent dye, researchers visualize how its distribution changes as the blade runs against a synthetic skin mimic. They can determine if the lubricant remains on the strip itself, if it creates a lubricating layer on the skin for the blades to glide over, and how it interacts with the skin during each razor pass. The lubricant formulation can be varied to see what provides the optimal lubricating film, or how the razor blade design and count influences lubrication.

The drawback of these optical techniques is that they require reflection and refraction of light and a laser detection system. This is not always possible or realistic due to space constraints, the optical properties of the materials involved, or the risk of a laser damaging the materials. Keen on making challenging measurements feasible, researchers at Imperial College developed a nonoptical approach to measure film thickness. Their approach uses sound waves. Ultrasonic waves have a frequency greater than 20 kilohertz, putting them outside the audible range of human hearing. These waves are commonly used in a variety of nondestructive testing and analysis techniques. Ultrasound is especially important for crack detection of machine parts. You want to ride that roller coaster with confidence that the track doesn't contain any micro cracks, right? Ultrasound makes it possible to inspect the tracks regularly without downtime. If a material is solid and has no separations caused by cracks, a sound wave propagates through it. If there's a crack, however, the sound wave will be disrupted.

Like light interference, the shift in the sound signal can be translated to distance traveled. When an ultrasonic wave encounters an interface between two different materials, part of the wave will be reflected off the interface. The portion reflected can be determined from a material property known as acoustic impedance, which represents the material's resistance to the propagation of ultrasonic waves and the speed at which sound travels through the material. When there is a gap in the interface, more of the sound wave is reflected. The amount is dependent on the size of the gap, the density of the medium filling the gap, and the speed of sound through that medium. If the gap is a result of a lubricant film, its thickness can be determined by knowing or measuring its density and the speed at which sound travels through it.

As a measurement technique, ultrasound has its limitations. It doesn't work with all materials. Polymers, for example, can give off noisy signals, making it difficult to parse the information. Ultrasound is, nonetheless, a powerful technique in tribologists' toolkit and has provided considerable insight into lubricated contacts and lubricant behavior. Sound-wave sensors can be hooked up to equipment operating in warehouses or wherever the application may be located to actively monitor the health of a lubrication system. Offshore equipment, for example, can be monitored to ensure that any seals or bearings maintain proper lubrication. If the length of the reflected wave changes, that can indicate that maintenance is necessary, preventing catastrophic failure. This real-time feedback can reduce downtime. Thurston would have been thrilled.[30]

All of these techniques took considerable effort and ingenuity to develop. The fact that so many scientists and engineers have focused on this particular area of tribology emphasizes just how important lubrication theory is. We saw a glimpse of this with transmission fluid, but in addition to machinery, lubrication regimes affect us in ways that may surprise you. The human body, composed of numerous joints and tribological contacts, is possibly the most important machine there is. Knees and hips are notorious for needing to be replaced as they wear down. Prosthetics must be designed for low friction and wear, but it's even better if we can avoid replacing them in the first place. To do so, we need to protect the longevity of our cartilage, the connective tissue at the end of our bones. As long as the cartilage is there, we're well-oiled machines.

Cartilage protects our bones from rubbing against each other when we use our joints. The problem is that cartilage degenerates with use and age. When that happens, we end up with bone-on-bone contact. That is painful enough, but now those bones are rubbing together and wearing out faster. As this happens, say hello to invasive surgery to implant an artificial joint. Like the majority of our body, cartilage is mostly water—up to 80 percent water, in fact—and proteins. Its biphasic nature makes it incredibly strong, able to handle high contact stresses as we move about. When the cartilage is subjected to loading, its fluid pushes against its collagen scaffold, and this pressure helps support the load.[31] The other function of cartilage is to reduce friction in our joints. The coefficient of friction for cartilage is orders of magnitude smaller than the famously slippery PTFE / Teflon. The synovial fluid that lubricates cartilage is made up of water, hyaluronic acid, and a variety of proteins, including those from our blood plasma. Hydrated cartilage is key to low friction, but fluid gets squeezed out during the loading process of the joint. This happens in your knees as you walk down a flight of stairs. Rehydration of the cartilage is critical, but sometimes cartilage doesn't rehydrate properly. This can happen as we age or as a result of certain chronic diseases like diabetes, high blood pressure, or arthritis. Without proper rehydration, the cartilage can't withstand the loads it should be able to handle, resulting in wear and tear of bones and tissue along with higher friction.

Lubrication regimes and fluid film behavior can offer key insights into cartilage health and hydration. Researchers have applied interferometry methods in the lab, simulating conditions of articular motion, that is, motion that employs the joints, such as walking, to see what happens to the lubricating fluid in cartilage—where the fluid is, whether it has formed a film, and how thick the film is. Cartilage spends a significant amount of time in the boundary lubrication regime. When the joints are under high load or moving slowly, cartilage surfaces may only be separated by one or two molecules of lubricating fluid. That's all the cartilage needs to maintain low friction. Designers of artificial joints optimize materials and surfaces to minimize friction in the boundary regime, running tribological tests under these conditions to iterate through designs.

It was long thought that cartilage only experiences boundary lubrication. But recently, using optical interferometry, researchers have found that hydrodynamic lubrication can be achieved through tribological rehydration. This is hydration that occurs through the sliding of our joints. The sliding creates hydrodynamic pressure, which causes the fluid to squeeze into the contacts and fully separate the cartilage with a thin film. This discovery has provided insights into how exercise can help rehydrate cartilage. Through tribological studies and modeling, we're learning what level of activity we should maintain to help keep our cartilage healthy and hydrated. These studies are also enhancing tribologists' understanding of the various mechanisms involved in cartilage lubrication, which may lead to therapeutics that can improve the properties of natural or engineered cartilage.[32] In cases where the cartilage has degenerated to such an extent that implants are needed, these studies can inform the development of optimal materials to mimic cartilage.[33]

Lubrication regimes are also important when it comes to another part that we use to mimic something in the body: contact lenses. If you suffer from myopia like me and approximately one billion others, you too may be wearing contact lenses right now. Contact lenses experience two different lubrication regimes during the course of a single blink. Blinking seems so simple, yet like most things involving our bodies, it is quite remarkable when examined closely.

We blink a lot. On average, we blink between twelve and twenty times per minute, which is somewhere between 14,000 and 19,200 times each day. If you're doing the math, that's between five and seven million times each year. Blinking on its own is a lubricating function. A blink refreshes our tear film, which protects the cornea. Contact lenses, which are made of polymers and water combined into a material known as a hydrogel, must correct our vision while allowing oxygen through to maintain a healthy eye and maintaining lubrication for a comfortable blink. They must also stay in place while this rapid-fire blinking motion is happening thousands of times a day. Easy, right?

Friction is an important indicator of the comfort of lenses in our eyes. The higher the friction, especially between the lens and cornea, the more uncomfortable the lens will be. Having proper lubrication to maintain low friction is crucial. It should be no surprise, then, that lubrication regimes play a critical role in contact lens comfort. Having a film of lubricant separating the lens from both the eyelid and the cornea provides the most comfort, but the act of blinking sends the lens through two different lubrication regimes. When our eyelids close during a blink, they move more rapidly than when they open. This speed difference, combined with the viscoelastic nature of the contact lens, changes the lubricant film behavior. The rapid velocity of the eyelid closing creates hydrodynamic lubrication, pushing the lubricant between the contact surfaces and creating a film thick enough to separate them. During the slower velocity of the eyelid opening, the film is thinner. This means there's a chance the eyelid will touch the contact lens, a prime source of discomfort. Meanwhile, the lens slides over the cornea at a much lower velocity compared to the speed of blinking, putting it in the boundary lubrication regime. Knowing this, different lens materials and lubrication formulas can be tested in the boundary regime to achieve the lowest friction possible between both the lens and the eyelid, and the lens and cornea. Hydrogels are used for their ability to meet a variety of needs. They allow the passage of oxygen to our eyes to keep them healthy, conform comfortably to our eyes, and are wettable and slippery.

There is some controversy in the tribology community over whether the Stribeck curve can be applied to soft biomaterials such as hydrogels. Lubrication theory is based on the behavior of machine components made of hard materials, such as steel. Studies of contact lenses seem to support the view that the Stribeck curve applies to these softer materials. But trust me when I say that shouting matches have erupted over whether it's truly a Stribeck curve. Semantics? Perhaps. Whatever your perspective on this debate may be, for softer and harder materials we can predict with confidence how thick the lubricant film is under various conditions and how friction changes. Using the different visualization tools available, such as the optical interferometry experiments previously discussed, tribologists are able to simulate the effects of blinking on a contact lens and measure the film thickness and coefficient of friction. By plotting friction dependence on load and speed, tribologists can confirm the shape of the curve and lubrication regimes and determine if the lens is in contact with the eyelid and cornea or if they are separated by a film.[34] All of these factors contribute to the comfort and functionality of the contact lens.

Sometimes lubrication involves emulsions, like yogurt or custard, which consist of at least one liquid dispersed in another liquid. The two liquids, like oil and water, don't usually mix together, but when shaken, they form an emulsion. How we experience food emulsions depends on how they lubricate the tongue and mouth. In food tribology studies, the aim of the game is to understand the processing of food, from oral lubrication to sensory perception. Just as friction can be an indicator of eye comfort or discomfort, so too can it determine our perception of food. If a yogurt is produced with lower fat milk, how will that change the mouthfeel? Researchers mimic the hardness, elasticity, and surface roughness of the tongue. They use a tribometer to measure the synthetic tongue material against a harder material that resembles the palate and add some saliva to the system. Between the tongue and palate is the food of interest, like yogurt. The tribometer runs at different loads and speeds, simulating various chewing conditions, recording the friction. To correlate this to user experience, food scientists conduct surveys to gauge people's sensory perception. Then, they connect the feedback to a desirable coefficient of friction. Often these tribological results are combined with rheological measurements, during which the viscoelastic properties of the food are measured as the sample is sheared. This allows scientists to understand how the breakdown of the food during chewing connects to friction and sensory perception during eating. Food scientists can then adjust formulations, adding thickeners or binders to improve mouthfeel and texture. This is especially important in foods where protein and fat sources are substituted, such as in plant-based yogurts and vegan meats.

Some studies have even suggested there are additional regimes for complex, heterogeneous materials like yogurt.[35] These so-called frictional zones represent the various stages of eating, from the liquid entering the contacts to consumption. Initially, yogurt seeps into the contacts, forming a thin film where there is a small gap between the tongue and palate and the low sliding velocity of the mouth. Next, the gap between the surfaces widens as fat particles spread out across the contact areas, generating higher friction. As the sliding velocity increases, a thicker film forms, reducing friction until the layer becomes thick enough for drag to increase friction or for the structure of the yogurt to break down. The result doesn't look like a Stribeck curve, but it follows the same principle, varying parameters like speed and load to see how the lubricant film develops. Food tribology is one of the newest areas of friction work, and there is still a lot to learn. There isn't a clear consensus on how to apply it, as tribologists are still faced with the challenge of comparing their results to subjective measures, such as survey responses. Researchers just know that once they connect the dots between human perception and the coefficient of friction, food formulation will become a much simpler task.

Recently, the Stribeck curve has found use in another unexpected application. In Chapter 2, we saw how static friction affects plate tectonics and earthquakes. Fluids with different viscosities, such as water, mud, and methane can creep into faults, potentially influencing the motion and stability of the plates. During her doctoral studies at École polytechnique fédérale de Lausanne, Chiara Cornelio was studying how the viscosity of fluids between plates affects lubrication regimes, friction, and plate stability.[36] Using samples of marble and granite slabs to mimic porous rock plates, she added liquids of different viscosities between the slabs. Then she measured the friction and stress of the slabs as they slid over each other. Cornelio found that the fluids followed EHL theory and that higher viscosity fluids create a lubricant film that reduces friction, stabilizing the fault between plates. The results of the study may deepen our understanding of how human practices such as drilling and fracking, which alter the viscosity of fluids between faults, influence seismic activity.

Drag

Earlier we saw that friction begins to increase after reaching a minimum at the onset of hydrodynamic lubrication. This may be caused by drag, the force that opposes motion involving a fluid. Sounds familiar, right? If you're wondering why drag isn't just called friction, that's a fair question. Different people will answer it in different ways. To some, friction is strictly the interaction between solid materials. For others, it's all friction, and drag is just another term for the kind that involves fluids. In this book, we'll subscribe to the view that drag is a type of friction involving fluids. It can occur between layers of fluid, or between a fluid and a solid surface.

Given that viscosity is also resistance to motion, you wouldn't be wrong to assume that viscosity and drag are connected. However, it would be a mistake to assume that viscosity *is* drag. Viscosity is a material property; drag is a force. Viscosity can, and does, influence drag, but drag can also stem from sources other than viscosity, such as the properties of the surface the fluid is moving over. This of course means that there are different types of drag. If you guessed that the type of drag induced by the viscosity of a fluid is called viscous drag, you are clearly catching on to the naming schemes of tribologists. *Viscous drag* is velocity dependent, but if the complexities of viscosity have taught us anything so far, it's that it's not as simple as saying there's a direct proportionality between the two. Viscous drag depends on the fluid and whether the flow is laminar or turbulent. At low flow speeds, typically in laminar flow, viscous drag is directly proportional to velocity. At high speeds, and in turbulent flow, drag is proportional to the velocity squared.

Viscous drag is also called skin friction drag since it is the result of the viscosity of a fluid as it flows along a surface. Familiar examples of viscous drag are ships or our own bodies moving through water. Altering roughness, applying coatings, and shifting the lubrication regimes are all useful tools in reducing viscous drag. The Speedo LZR swimsuit, now banned at the Olympics, was excellent at reducing viscous drag. The fabric contained fewer stitches thanks to ultrasonically welded seams, providing a smoother surface for the water to flow over.

In our bodies, blood experiences viscous drag from the walls of arteries. Variations along the wall of the artery, such as curves or white blood cells adhering to it, alter the viscous drag of our blood, impacting blood flow.[37] Atherosclerosis is a condition that occurs when plaque builds up on the arterial wall, increasing viscous drag and requiring the heart to work harder to pump blood. It is believed to affect up to half of adults between the ages of forty-five and eighty-four.[38] Atherosclerotic plaque formation is found in the inner curvatures of coronary arteries, where laminar flow has been disrupted and transitioned to the faster turbulent flow. In one experiment, scientists found that plaque formation was a result of reduced drag forces on the arterial wall. This may sound counterintuitive since drag forces increase with speed. But in turbulent flow, as the fluid creates vortices, its molecules generate drag forces in many different directions. This means the artery wall experiences less drag under turbulent flow than the more streamlined laminar flow. Without enough drag to shear plaque off the artery wall, it accumulates, causing blockages. With this knowledge, stents can be designed to ensure laminar flow and inhibit plaque buildup.[39]

Understanding viscous drag is also critical for plate tectonics. Mantle convection, a heat transfer process, is central to plate tectonic motion. The mantle, the mostly solid layer of Earth's interior, is heated from below by the hot core, and this heat moves to the surface of Earth through convective currents. These currents cause motion in the plates at the crust as warmer material rises. Three forces, not necessarily independent of each other, determine how the plates move: the weight of the plates themselves, the buoyancy of the mantle rising, and viscous drag. In this case, viscous drag is the force opposing motion of the plate against the mantle moving beneath it. This resistance can either drive plate motion or fight against it.[40] We saw how viscosity can influence the sliding stability of Earth's plates. Viscosity changes in the mantle may also influence tectonic movement through viscous drag. Recently, researchers realized that so-called low-velocity zones in the upper mantle, where seismic waves move more slowly, may be more abundant than previously thought.[41] This is believed to be a result of partial melting of the rocks and minerals in the mantle. Teams of researchers are beginning to figure out the viscosity of these zones based on the motion of tectonic plates. In 2018, researchers at the University of Chicago studying a deep ocean earthquake discovered that the motion they measured would only be possible with a rela-

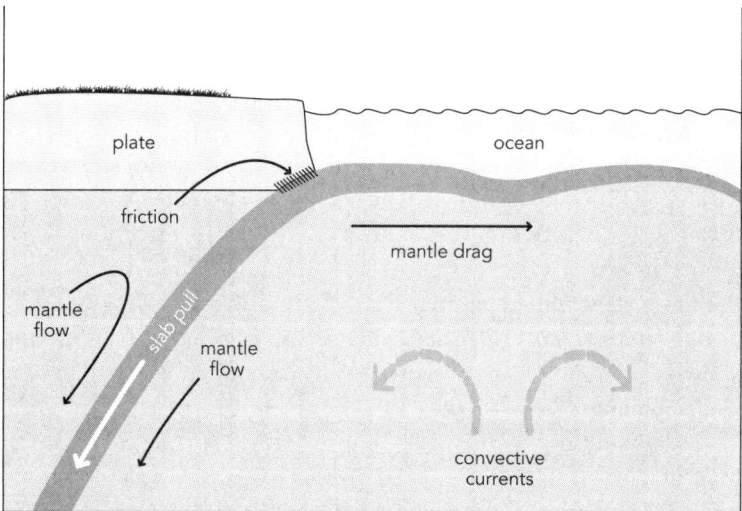

Forces involved in the motion of Earth's plates

tively thin low-viscosity layer between the upper and lower mantle.[42] They believe that understanding this low-viscosity layer will help improve models of tectonic movement and the convective currents that contribute to motion, viscous drag, and ultimately the earthquakes we experience topside.

Gases as Lubricants

So far, the examples of fluids in this chapter have all involved liquids. But gas is also a fluid and can be used as a lubricant. One of the more common applications of gas lubricants is air bearings. Used in applications ranging from space simulators and nanosatellites to dental drills and disk drives, air bearings are similar in function and appearance to conventional bearings but use a thin layer of pressurized gas to separate the bearing and shaft surfaces during operation.

Air bearings were developed by Thurston's pupil, Alfred Kingsbury. Following in his teacher's footsteps, Kingsbury wanted to increase the load-bearing capability of equipment while decreasing its friction. Working with a horizontal piston assembly, he noticed that the spinning piston never touched the cylinder wall and yielded low friction during operation. He realized that a layer of air must be forming between the piston barrel and cylinder supporting the weight of the horizontal piston. If he could figure out the mechanism behind this air layer formation, he could design machinery to generate it during operation.[43]

Kingsbury found that the air pressure separating the piston barrel and cylinder was generated by the rotation of the piston. If the system was unable to spin at a high-enough velocity to provide the necessary pressure, the air layer would disappear. Today, an injector system can supply gas at pressures that separate surfaces with a thin gas film. There are also hybrid designs, which rely on external gas supplies at low velocity and then switch the gas off at higher velocities. One of the benefits of air bearings, or similar systems that rely on a layer of air to separate surfaces, is that the only friction you need to worry about is viscous drag between the gas and solid parts. As we've seen, this friction is lower than friction from solid-on-solid contact. What's more, solid-on-solid contact is avoided without the use of oils and greases, making maintenance easier. Operators don't have to schedule checks and refills to ensure there's enough oil or grease present. There tends to be less wear since surfaces don't interact with chemicals or particulates in the greases, and parts last longer. And there's less environmental contamination. This is particularly important for applications such as food and pharmaceuticals, for which you don't want to risk any contaminants sneaking into the product.

Air lubrication is also being used to tackle one of the biggest nuisances in train travel: the friction between the railway and wheels. In this case, the nuisance is that the friction between the wheel and track is too *low*. In England, a common grumble from commuters, particularly in the autumn, is the infamous "leaves on the line" delay. How do leaves on a rail line lead to delays? Surely a train can just push them aside and carry on with its business. Unfortunately, that's not the case. The combination of wet weather and trains running over leaves repeatedly turns them into a paste that causes trains to skid, all because of low friction.

After vexing riders for decades, tribologists have begun tackling this problem. It turns out that leaves on rail lines have the same coefficient of friction as skates on ice. Even as a tribologist, I find that remarkable. They don't call leaves on the line the black ice of rails for nothing! In fact, the coefficient of friction for the film left by wet sycamore leaves is even lower than that of pure water.[44] Biomolecules in the leaves called tannins react with the iron in the steel tracks, producing an ultra-low-friction film that can wreak havoc on rail travel. As heavy trains roll over the biofilm, it becomes compact and viscous. The viscous material acts like a lubricant film, separating the wheel and rail and reducing the traction of the train. Of course, this naturally leads tribologists to other questions. For instance, do different leaves produce a different type of biofilm, and will that biofilm have different tribological behavior? What causes the initial chemical reaction, and how do we prevent it in the first place? Why is the biofilm so difficult to remove? Does frictional heating from the wheels catalyze the reaction or strengthen the bonds between the track and biofilm? And hearkening back to Coulomb's concerns in Chapter 1, if we're recreating the problem in a lab, how do these idealized lab experiments compare to the real-world scenario?

To answer these questions, researchers at the University of Sheffield have been using actual train car wheels on cuts of tracks in their lab, instrumenting them to act as tribometers.[45] As a result of these experiments, Britain's railroad system, Network Rail, has employed a variety of solutions to mitigate the issue, including special leaf-clearing trains that spray high-pressure water on the line to break up the films and then apply a sandy coating to improve adhesion between the wheel and rail. Researchers are also looking into the tannin levels of different trees and assessing whether that influences the film formation and friction. If so, an assessment of the types of trees along the lines could indicate where heavier treatment is needed, or if other trees should be planted in those areas. Rail companies have even considered applying dry ice, which can freeze the slicked down leaves, causing them to expand upon thawing and making them easier to remove. Tribologists continue to develop coatings and treatments that may help reduce the chemical reaction that produces this slippery film.

The ultimate solution, however, comes from air lubrication, which removes the contact between the wheel and rail all together. Maglev trains, those levitating trains that seem to be straight out of a science fiction novel, do exactly that. These trains, found in China, Japan, and Korea, rely on magnetic forces to create an air lubricant gap between the train and a concrete guideway. Superconducting magnets, placed on the undercarriage of the train and on the guideway such that their like poles face each other, provide repulsion for levitation. Loops along the tracks produce electrical currents, which are used to magnetize electromagnets along the track and train. The magnetic force supplied by electromagnetic materials is temporary, so these currents act as switches, turning the forces on and off. These switches then push and pull the magnets on the train to propel it forward. Maglev trains, with the few inches of air they float on, can reach velocities well over 300 mph.

Lubricating in Fluid Environments

As beneficial as gas can be, it comes with its own set of challenges. Ultimately, gas is a tricky lubricant due to its volatility. Gases also don't have the cooling capabilities of liquid lubricants, whose higher heat capacity enables them to absorb more heat before their temperatures rise. This makes overheating a concern. Even in environments where overheating isn't an issue, such as the ocean, gas can prove difficult to use as a lubricant—it's hard to maintain a stable gas film.

Some engineers have been applying air lubrication to ships to reduce drag. This approach, which involves coating the bottom of a ship with tiny bubbles, less than a millimeter in diameter, highlights the challenges of air lubrication. Bubbles should, in theory, reduce viscous drag as water runs against the bubble coating instead of the solid hull. It's easier for a ship to move through the less-dense air bubbles than through water. The problem is that bubbles pop. To maintain a coating on the hull, there needs to be a steady supply of bubbles. Some ships are equipped with a bubble injection system that pressurizes air, sending it to the ship's hull, where the pressure difference between air and water creates bubbles. Large, flat-hulled, and frankly inefficient vessels like cruise ships are the best candidates for this technique. V-shaped hulls will experience less benefit as the bubbles are more prone to floating away. The speed of the vessel also matters. If the ship slows down or experiences rough seas, the bubble coating can be disrupted as bubbles pop and move around. Bubble coatings generated a lot of excitement when first introduced, but the fanfare has died down, in part because it has taken about two decades to gauge the effectiveness of this approach. As you can imagine, quantifying its impact is a bit of a nightmare. That's because the overall drag on a ship is influenced by multiple variables, including viscosity, speed, and ship design. Some researchers believe it's possible to reduce the overall drag of a ship by up to 30 percent with a bubble coating. Others are more skeptical. But with the potential energy savings available, it's worth pursuing bubble coatings.[46]

In recent years, another approach to drag reduction, superhydrophobic surfaces, has renewed excitement in air lubrication. Superhydrophobic materials really don't like water and repel it incredibly well. Molecules that are hydrophobic are generally nonpolar, meaning there is no charge distribution across the molecule. You've seen hydrophobic materials before, whenever you've noticed beads of water form on a leaf. These materials, by repelling water away from their surface, could potentially provide another means of generating bubble coatings by creating a thin air layer between the surface and water. Research has gone back and forth on the feasibility of superhydrophobic coatings, due in large part to the complexity of the overall problem. As is so often the case, recreating boating conditions in the lab is incredibly challenging. A great air lubrication layer may form in a lab flow test that just doesn't materialize or stick around in the real world. Cost and ease of manufacturing are also a concern.

As we know from the lambda ratio, surface roughness is a key component of lubricant thickness. It's particularly important for superhydrophobic materials because roughness can enhance hydrophobicity. A rough surface has more surface area than a smooth one, making it harder for water to stick to it. This can be seen with beads of water, which become more spherical on rougher surfaces.[47] If micro surface roughness can be controlled, a layer of small beads may be able to form.[48] For this to be feasible, methods need to be developed to produce superhydrophobic materials more economically. There's also a complication: some researchers have shown that hydrophobic coatings may actually *increase* drag. With so many different variables at play, this remains an active field of research. Of course, there's the design of the boat itself and the fact that skin friction, or viscous drag, isn't the only kind of drag acting on the vessel.[49]

Dragging On

Another significant source of drag on water-faring vessels is *form drag*. Form drag is also referred to as pressure drag because it stems from the pressure of a fluid pushing against an object as it tries to move through the fluid. Whereas viscous drag occurs in the direction of the vessel's motion and stems from friction between the fluid and surface of the object, form drag occurs normal to the direction of motion and stems from the size and shape of the vessel. Picture a stick floating in water. It's easier for it to move downstream when it's parallel to the flow of water as opposed to perpendicular to it. This is because water can more easily flow around it, and form drag is low. When the stick is perpendicular to the flow, the water has to work harder to move around it. You may see a small wave of water swell at the front of the stick, followed by the formation of vortices along the ends as water finally flows behind it. Much like the stick, the shape of a boat will affect how much form drag is experienced.

Form drag is why ships made for speed, like the Viking longships, have a very different shape, both above and below water, than a large freight vessel or cruise ship. A larger cross section, such as one with a wider hull and more surface area perpendicular to the direction of travel, results in higher form drag. Drag scales exponentially with speed. A boat whose speed doubles will experience four times as much form drag. That's why those zippy high-speed boats have such a streamlined shape. Form drag is often the dominant type of drag, which is why so many researchers work in labs with flow tunnels, experimental facilities where they can submerge objects and study the resulting fluid dynamics. This enables them to test designs, as well as to model flow to create the most efficiently shaped vessels for the job at hand.

You've probably heard this kind of streamlined design referred to as aerodynamic. That's because fluids aren't just liquid. Boats have to move through both types of fluids: liquids and gases. As the bottom of the boat cuts through water, the top of the boat experiences viscous and form drag from air molecules instead of water. Truly aerodynamic vessels, such as the sailboats racing in the famous America's Cup, sit mostly above the water, slicing through it on ski-like airfoils to minimize drag from water.

Aircraft, while they don't have to deal with water, experience both viscous and form drag—viscous drag from the friction between air molecules and the surface of the aircraft and form drag from the shape of their design. Just as with boats, the shape of the aircraft influences how the fluid, in this case air, flows around the body, and this flow affects drag. Unlike ships in the sea, however, aircraft face additional drag from *lift,* the force holding a plane in the air and perpendicular to the airflow. Today, thanks to Wilbur and Orville Wright's historic flight in Kitty Hawk, North Carolina, in 1903, aircraft wings are carefully engineered to reduce this component of drag and increase efficiency.[50]

Understanding how lift influences drag proved key to achieving manned flight. Flight was not a new concept at the time. Through the Wright brothers' notebooks, we know that they were successful because of the aviation giants who came before them. In 1889, the German engineer Otto Lilienthal, often referred to as the first aviator, explained lift in his book *Bird Flight as a Basis of Aviation.* The book was so influential in the development of aviation that it is still in print over a century after its initial publication. In it, Otto methodically documented experiments in what can be considered an early wind tunnel.

Born in 1848 in Anklam, Prussia, Lilienthal, or the Flying Man as he would eventually become known, was one of three children. He credited both of his parents with fostering his and his brother Gustav's mechanical inclinations. Fascinated by flight and an engineer to the core, Lilienthal began observing storks near his home. Noticing that storks would take flight by charging into the wind, he surmised that it must be easier to rise against the wind than with it.

The Lilienthal brothers tested their hypothesis with a two-by-one meter wing of beechwood. They would run down a hill to gain speed, then leap into the air against the wind and try rising up like a stork. Had five-year-old me known that famous scientists had done this, perhaps I could have explained my own decision to conduct a similar experiment—with an umbrella and the top of the garden wall, much to my mother's horror. Like me, the Lilienthal brothers didn't achieve flight on this attempt, but they became determined to understand aviation. Otto went on to study mechanical engineering at what is today the Technical University of Berlin where he continued his investigations into flight. He realized that an important initial step was to be able to control gliding. This meant the brothers needed to contend with friction.[51] By this point, it was known that friction acted on objects moving through air. Otto Lilienthal grasped the importance of drag as it pertained to flight.

One of the brothers' experiments involved studying the mechanics of a whirling arm, a centrifuge-like apparatus that consisted of a rotating arm suspended on a balance. By adjusting the weight and speed of the arm, they could measure lift by how off balance the arm became and drag by the time it took for the arm to rotate around the device. As they varied the shape of the arm, they meticulously documented the results, creating the first table and plots of the lift-to-drag ratio. The lift-to-drag ratio, remains one of the most important numbers in airfoil design. If the drag is higher than the lift, the object attempting to fly will slow down, or worse, not fly at all. The goal is to maximize this ratio to enable lift with the least amount of resistance.[52] Otto measured the lift-to-drag ratio of flat and cambered airfoils. Cambered airfoils are the wing shape used in today's aircraft. These airfoils are designed such that the top surface has more curvature than the bottom. This causes air to move faster under the wing than over it, creating more pressure from below and enabling lift. While Otto was on the right track, the rotary nature of his whirling arm and its limited speed range meant that the brothers didn't see appreciable differences between the two designs. In fact, Otto even believed, incorrectly, that flat airfoils could possibly work. Otto used these findings to design gliders but was not convinced powered flight was possible. Tragically, he died from a fatal gliding accident and never saw his dream of powered flight realized.

Despite not recognizing the value of cambered airfoils, Otto Lilienthal's work contributed to the success of aviation in multiple ways. First, he approached design as a series of controlled experiments. Second, he broke down the air force into its lift and drag components. In his own words, "All flight is based upon producing air pressure, all flight energy consists in overcoming air pressure."[53] Perhaps most important, he published and lectured on his findings. As a result, his work found its way to the Wright brothers in Ohio, who would finally unlock the way friction was holding back flight.

At the heart of the Wright brothers' glider design was the belief that they needed to control the stability of the aircraft through the aerodynamics of the wings. Gliders like Lilienthal's relied on the human body to shift weight and control the direction of motion. This is impractical for powered flight, so the Wright brothers set out to achieve aerodynamic balance, focusing their efforts on the wings. Applying the tables Lilienthal had produced, they designed wings large enough to withstand the gusts of wind they expected to experience. Despite this, they found themselves unable to control the gliders and struggled to achieve lift. The actual lift was only a third of what they had predicted based on Lilienthal's tables. On top of that, the drag was higher than expected, resulting in a lift-to-drag ratio that was too low to achieve gliding. It was the classic problem of friction offering too much resistance to motion and, in this case, preventing flight. The question, of course, was where the miscalculation was.

The Wright brothers spent the winter of late 1901 through early 1902 designing a wind tunnel to measure lift and drag, then conducting experiments to replicate the values Lilienthal had previously calculated in the lift-to-drag ratio. The tunnel consisted of a wooden box about six feet long with a simple fan blowing air down it.[54] The fan speed was controlled by a gear and pulley system and could provide winds of up to twenty-five miles per hour. Inside, the brothers placed a balance to hold the airfoils. Based on the angles of the balance, they were able to work out the lift and drag forces for each airfoil design. Wilbur Wright's diary entries from this period offer insight into this process, documenting their results, flights at Kitty Hawk, designs, and scientific musings. With their new experimental data in hand, they concluded that Otto Lilienthal's tables were incorrect because the measurements were specific to one airfoil shape. As a result, his drag measurements hadn't accounted for all the different types of drag forces generated by the shapes the Wright brothers were using.

One such force was induced drag, a type of air friction that stems from the flow of air over the wing tips. For flight to occur, pressure must be greater under the wing than above it, generating lift. But this pressure differential also causes air to flow up from underneath the wing's edges and mix with lower-pressure air, creating vortices. These vortices deflect flow downward at the trailing edge of the wing. Lift, however, is always perpendicular to the flow of air, not the wing itself. As a result, the lift of the wing is now at an angle offset from vertical. The component of this angled lift that is parallel to the direction of motion is the induced drag. Unlike most drag, *induced drag* is inversely proportional to the square of speed. At high speeds, lift can be generated with a small angle of attack between the wing and airflow. The reduced angle of attack diminishes the strength of the vortices and induced drag at the edges of the wings.

The Wright brothers solved the problem of inadequate lift due to induced drag with a cambered wing design. A well-designed cambered airfoil reduces the friction at the wing tips by minimizing the strength of the vortices, enabling the available lift to overcome drag. But induced drag was just one of the challenges caused by drag. As the aircraft turned, the air flowed over and around the wings, creating drag that was not symmetric on both sides of the plane. This asymmetry could swing the aircraft opposite where the brothers intended it to turn. This time, the solution wasn't in the wing design but at the tail of the aircraft. The brothers' glider had a fixed tail. They realized a rudder could resolve this issue, enabling the pilot to direct the airflow via the tail to stabilize the craft. In 1902, they filed a patent application for their movable tail.[55] With this invention, their glider became airborne and, crucially, controllable.

Today, many commercial aircraft are equipped with winglets, vertical tips at the end of the aircraft's wings. Winglets reduce the induced drag at the tips of wings by smoothing the airflow across the wing and minimizing vortices. The energy that would be lost to the vortices can instead be used to propel the aircraft forward. The size and shape of winglets varies considerably, depending on the aircraft size and typical speeds. Some curl up; others are flat. This friction-reducing innovation has spawned its own industry, with manufacturers specializing in retrofitting aircraft with winglets.[56]

We want our aircraft to go farther and faster—and to do so more efficiently. One of the aims of aircraft design is to maximize the lift-to-drag ratio by minimizing viscous drag.[57] Engineers use a variety of techniques to manipulate viscous drag. First, they design the most streamlined and aerodynamic body possible. The shape of the wings with respect to the direction of airflow is designed to minimize turbulent flow, viscous drag, and lift-induced drag. This is achieved with the cambered design and by ensuring edges are rounded to enable smooth flow. Second, they ensure the body of the plane is as smooth as possible. Rivets and fasteners are as flush as possible and exposed surfaces are treated with paints, coatings, and polishes to reduce viscous drag. Recent materials have included designs inspired by nature, such as coatings that mimic shark and dolphin skin. The microstructures of these skins align the fluid flow over the body such that it minimizes turbulence and reduces drag.[58] Some of these coatings are in early stages of development, highlighting the ongoing efforts of scientists to manipulate fluid friction in aviation.

Fluid Friction and Weather

I'd be remiss if I concluded this chapter without addressing one of the most significant ways in which fluid friction, including drag, influences our lives: the weather. Recently, bigger, more powerful hurricanes have been in the spotlight. As anyone who lives in the coastal southeastern part of the United States knows, hurricanes in late summer and early fall are as expected as winter snowstorms in other parts of the country. Hurricanes reaching Category 3 or higher are considered major hurricanes. In recent years, we've seen considerably more Category 4 and even Category 5 hurricanes. A twenty-four-hour news cycle and streaming content make information about these mighty spectacles of nature instantly available. Friction has become a buzzword in tropical meteorology coverage. This has caused confusion about the role of friction in the weather. Warming oceans and climate change are linked to the increase in hurricane strength, so people have asked me if the friction meteorologists talk about is also somehow related to climate change.

This, as it turns out, is a complicated question. Friction is an important player in hurricane winds, and friction is generated by winds that can be affected by climate change. Winds are determined by a combination of forces. Two of the most significant factors are pressure gradients and the Coriolis effect. As we discussed with lift, air moves from areas of high to low pressure to seek equilibrium. The change in pressure over distance is called a pressure gradient, and the stronger the gradient, the more powerful the winds will be. The Coriolis effect determines the direction of winds as they move from high to low pressure. It is the reason why winds will blow counterclockwise around a hurricane in the Northern Hemisphere, but a similar system rotates clockwise in the Southern Hemisphere.

Earth's poles are areas of high pressure, whereas the equator is an area of low pressure. If Earth wasn't rotating, air would simply circulate between the poles and equator. But Earth is rotating. At the equator, Earth's widest point, almost 25,000 miles must rotate in a twenty-four-hour period, meaning equatorial areas are moving at around 1,038 mph. The poles, on the other hand, are moving at only a little more than 3 inches per hour. When viewed from the North Pole, Earth rotates counterclockwise. If you try throwing something from the equator to someone in, say, Greenland, the object will land to the right of them because they're moving slower and haven't caught up. If instead you throw something from Greenland to the equator, the object will again land to the person's right because they're moving faster than you and are ahead of the object. Earth's rotation, when viewed from the South Pole, is clockwise, so if you repeat this exercise in the Southern Hemisphere from the equator to, say, someone in Chile, the object will land to your left. This deflection is the Coriolis effect, and circulating air experiences it, too. Weather patterns can be predicted based on the Coriolis effect and pressure gradients, but highly sensitive predictions based on just these two effects alone will come up short. That's because friction comes into play at Earth's surface.

When air moves across the surface of Earth, there's resistance to the movement. The magnitude of the resistance depends on the speed of the air and the surface in question.

If this sounds familiar, that's because our understanding of fluid dynamics and fluid friction applies to weather too, except now we're looking at much greater areas than aircraft wings in the air or Viking ships cutting through the ocean. Just as an object in the sea causes water to flow around and over it, so too does the topography of Earth. The friction between the surface and the fluid, which in this case is air, affects how strong the flow is and the direction it takes. Friction disrupts airflow, making it more turbulent than it would be if the air were flowing across a uniform surface and creating gusts. This is especially the case in the first few thousand feet of the atmosphere, known as the *friction layer* or planetary boundary layer. In this layer, surface friction influences wind speeds. As you move farther up in the atmosphere, the influence of the friction layer diminishes. The height of the friction layer isn't constant and depends on the topography of the terrain. The friction layer extends farther into the atmosphere for rougher terrain, where surface friction is higher. The disruption to airflow caused by friction reduces the Coriolis effect on the wind, and the pressure gradient dominates wind direction. The wind moves at an angle toward lower pressure regions until forces from pressure, friction, and the Coriolis effect are balanced.

Velocity also affects surface friction. When air is still, or moving very slowly, friction is negligible. However, as air molecules bounce around in stronger winds, there's more resistance to their motion. This is why friction is mentioned so often during hurricanes whose wind speeds can surpass 150 mph. At such high speeds, friction causes noticeable shifts in air movement. In 2017, wind maps of Hurricane Irma, one of the strongest hurricanes on record in the Atlantic basin, demonstrated how wind speeds varied across the state of Florida as the topography changed. This wasn't just at the center of the storm, where winds are the strongest, but across the entire wind field of the storm. Winds were about 10–20 mph stronger over water than land.[59] While on the one hand, it may seem like friction is doing everyone a favor by slowing down winds, that same surface friction over land can have devastating effects. In the friction layer, it can cause such a gradient in the already powerful wind speeds that air will start to spin. When this spinning air intersects with a current of rising air generated by the hurricane, a tornado forms.

People are quick to criticize meteorologists for poor forecasting or overhyping of the storm when winds over land are weaker than projected. These responses are unfair. Friction is to blame. Sustained winds can drop by as much as 20 percent over land due to friction.[60] The recent trendiness of friction in weather lingo may make it seem like friction has only recently been recognized as an issue in hurricanes, but this isn't true. Meteorologists have been exploring the role of friction in hurricanes since at least the 1950s. In August 1949, a hurricane swept across southern Florida, over Lake Okeechobee. Unlike previous storms, this one passed over an area where extensive instrumentation had been set up by the Army Corps of Engineers. The surface wind speeds over land and the lake could be compared to see how friction affected the different topographies. The results showed exactly what the wind maps of Irma showed us over seventy years later: there was a significant difference in the wind speeds over land and the lake.[61]

Hurricanes are enormously complex systems, and friction is just one puzzle piece in accurately modeling and forecasting them. But as anyone who has tried putting together a thousand-piece jigsaw puzzle knows, it's not solved with only 999 pieces. Solving the puzzle requires understanding dynamic and variable conditions, including broader weather patterns, ocean temperatures, how the ocean and atmosphere interact, and atmospheric conditions such as pressure, temperature, and moisture.

Satellites and radar can offer an overview of what a storm is doing, but they can't provide critical data needed to forecast the storm, including tracking its strength. Factors such as the ocean temperature correlated to depth and atmospheric conditions are key to predicting storm behavior. For anyone who has ever dealt with a hurricane barreling toward them, or even just followed the tracking, you know that these storms can start to shift, turn, wobble, and strengthen in ways that models sometimes don't predict. That's because these storms are so dynamic and influenced by factors that atmospheric satellites can't pick up, such as the ocean temperature below the surface. Storms may also undergo rapid intensification, where winds increase by thirty-five miles over the course of a day, as happened with recent storms Michael in 2018, Ida in 2021, and Milton in 2024. To capture what's happening as a storm is intensifying so quickly, we need to drop instrumentation directly into it to make measurements of wind speeds and pressures.

Allow me to introduce the revered hurricane hunters. These planes, either from the 53rd Weather Reconnaissance Squadron of the Air Force or the NOAA (National Oceanic and Atmospheric Administration) Aircraft Operations, routinely fly into tropical weather systems in the mid-Atlantic, Caribbean, and Gulf of Mexico, making in situ measurements. I enjoy checking flight tracking sites to watch the hurricane hunters zigzag around the part of the map that every other flight is avoiding. Hurricane hunters generally cut diagonally across a storm, then make a series of left-hand turns to avoid cutting into the brunt of the winds, which rotate counterclockwise at these locations because of the Coriolis effect. If the hurricane hunters were to fly into storms in the southern hemisphere, where winds rotate in the clockwise direction, they'd instead make right-hand turns.

Hurricane hunters are equipped with state-of-the-art instrumentation that collects data on temperature, dew point, altitude, air pressure, and wind speed. But these planes, impressive as they are, can't fly close to the friction layer. To make those important surface measurements, they release dropsondes, tubes of instrument packages with small parachutes that report back as they fall to the ocean below. Dropsondes record temperature, pressure, and humidity, but not wind. An extrapolation method, based on the wind speed at the plane's altitude and the dropsonde's GPS latitude and longitude location, determines the wind speeds at the surface.

Drones are increasingly being used in meteorological data collection. NOAA employs a fleet of aptly named Saildrones, autonomous drones outfitted with an impressive array of sensors to gather information on the wind, air, and ocean. In 2021, the first drone sailed into the path of a hurricane, right through the eye of a Category 4 storm.[62] Not only did it record real-time data that was used for forecast models, but it made a video of its voyage through the rough seas, capturing stunning fifty-foot waves.

It should be noted, of course, that while this discussion has focused on surface friction, weather systems are governed by complex fluid dynamics, and surface friction is just one parameter forecast models have to consider. The most commonly used hurricane models, the Global Forecast System (GFS) and the European Centre for Medium-Range Weather Forecast (ECMWF, referred to often as the Euro), have parameters dedicated to various types of friction.[63] You may wonder why so much money, time, and effort is spent globally on a weather event that is localized to a small corner of the world. First, hurricanes may affect a small pocket of the world, but tropical cyclones are not limited to the Atlantic and Gulf of Mexico. They form in the Pacific as well. Each year, over half a billion people experience these weather events.[64] Second, insights gained from modeling storms can help with weather modeling beyond hurricane season. And third, while they may seem to affect a small pocket of the world, these storms are devastating, causing billions of dollars of destruction and loss of life.

Of course, not all this destruction is from wind. Storm surges can lead to fatal flooding and leave a path of destruction in their wake. Depending on the storm, much of the damage may come from the rush of rising water that can accompany a hurricane. Hurricane Ian, which struck Florida in 2022, had a devastating storm surge that accounted for nearly 150 deaths.[65] A variety of factors influence the extent of a storm surge, including the winds of the storm, the astronomical tides, the rainfall and runoff, and the surface roughness of the shore and ocean floor.

Nature has designed its own protection systems for many land masses. Barrier islands buffet the mainland of many coastal regions, bearing the brunt of a storm surge. They weaken the rushing onset of water to protect the mainland while also providing surface friction to reduce wind speed. Barrier islands are brilliant natural defenses.

They're also highly desirable places to live with easy access to the beach. Humans have adapted to the reality of living on barrier islands, building with hurricane winds in mind. Storm surges, however, remain the primary concern for inhabitants. The severity of storm surges can be as challenging to predict as hurricane strength. With any missed forecast, there's always the concern that people won't take the surge warnings seriously because of a handful of storms during which the surge wasn't as bad as predicted. I know this all too well, having grown up on a barrier island. This is why in Florida, researchers have been tasked by insurance companies and the Federal Emergency Management Agency (FEMA) with updating flood maps to indicate those areas prone to storm surges.[66] To do this, they mapped the roughness of the areas in question, both on land and offshore, and built in friction parameters to determine how storm surges are affected by roughness. The issue with such modeling is that surface roughness changes over time because of erosion and the disappearance of sand dunes.

Hindcasting, whereby researchers look at data after a storm has passed, is one way to understand this rate of change and its effects on flood zones. It has been incredibly effective in elucidating the role of friction in storm surges. The friction of water against the ocean floor is known as *bottom friction*. The rougher the seabed floor, the higher the bottom friction. Bottom friction is what causes waves to slow down as they rush to the shore. As the bottom of a wave interacts with the ocean floor, friction slows it down. Meanwhile, the top of the wave continues at a higher speed, resulting in the wave curling into itself and eventually breaking apart. In 2017 researchers hindcasting Hurricane Rita, which struck in 2005, found that lower bottom friction led to higher, faster storm surges.[67] Using water level, wind speed, and surge data collected during the storm, they incorporated various friction conditions into their model until their simulations matched the data from the storm. Their result highlighted the importance of knowing the magnitude of bottom friction. If a storm surge model fails to include or accurately capture bottom friction, the surge might be forecast to arrive earlier and be worse than it actually is. If this happens consistently, people in storm surge zones will stop taking warnings seriously, leading to increased risk of life. Measuring bottom friction during a storm is an ongoing challenge. It is difficult to directly measure the force along the current of water sweeping the ocean floor. Instead, it's typically extrapolated from wave height and terrain roughness. Hindcasting can help us understand the relationship between what we are able to measure and the magnitude of bottom friction. Hurricanes themselves will provide us the best data to do this, and with each storm, modeling improves.

As storms become more powerful, more people are finding themselves in the line of the storm surge. Hurricane Ian flooded sixty miles of Florida's coast. Some places saw water levels rise one or two stories. Such storms are only becoming a more frequent occurrence, as evidenced by the devastating flooding brought two years later by Hurricane Helene (2024). The size and power of storms has been attributed to warmer ocean temperatures, which provide them more energy, bringing climate change to the forefront of conversations during hurricane season. In theory, we could find ways to make the ocean floor rougher to increase bottom friction near shorelines and mitigate storm surges, but this would be disruptive to ocean life and, frankly, impractical. For now, the most promising solution is to address the human activities that are fueling climate change, beginning with our insatiable energy needs.

5 *A Waste of Energy*

ASK A TRIBOLOGIST WHY she does what she does, and you're sure to receive some amusing answers. A common refrain I've heard, and admittedly used myself, borrows a line from the mountaineer George Mallory. When asked why he wanted to climb Mount Everest, Mallory responded, "Because it's there." We study friction because it's there, and more specifically, because it's pretty much everywhere. Ultimately, scientists and engineers seek to understand the world around them and use that understanding to innovate and contribute to society. Why wouldn't a curious engineer or scientist want to study something as ubiquitous as friction? The impacts of friction are as far reaching as its applications. As we've seen, friction is the dissipation of energy. It shouldn't surprise us, then, that friction can help us tackle one of the most urgent issues we face: the energy crisis and climate change.

As Earth's temperature rises, weather events have become more extreme. Every year, it seems, storm strength, wind speeds, high temperatures, and number and intensity of wildfires reach new records. Antarctica is experiencing unprecedented warming. Its collapsing ice shelves are contributing to sea level rise, devastating low-lying lands and contaminating freshwater supplies. Across the globe, people are feeling the effects of rising temperatures in places that never used to experience heat waves. Were Earth's temperature to increase by even half a degree, the resulting evaporation would dramatically alter the atmosphere, amplifying warming caused by greenhouse gases and intensifying already devastating drought and flooding. Curbing global warming—by reducing our reliance on fossil fuels—while ensuring that we meet our energy production and consumption needs must be an all-hands-on-deck effort. But there's no single solution; rather, we must take a multipronged approach.

From the first fire millions of years ago, to the nineteenth-century work of Robert Henry Thurston, who connected friction to energy efficiency and waste, to Peter Jost's formalization of the field of tribology with his groundbreaking report in the 1960s on the economic impact of energy savings, the story of tribology has been deeply entwined with our expanding energy and efficiency needs. Since the Industrial Revolution, global energy consumption has risen to around 600 exajoules a year.[1] One joule is the energy of one newton of force applied over one meter, which is about the same amount of energy required to lift a tomato one meter in the air or flip on a single Christmas light for one second. An exajoule is one joule times 10^{18}, which is roughly equivalent to the energy produced by 174 million barrels of oil. That's a lot of energy. Fortunately, friction is a powerful tool in helping us reduce energy consumption.

In 2016 a group of scientists and engineers gathered in a conference room in a large Las Vegas convention center to discuss how tribological solutions could help the US mitigate the energy crisis. Unlike that 1964 conference in Cardiff, Wales (mentioned in Chapter 1), this time everyone in the room was a tribologist. Instead of trying to convince others of the need to study and understand friction, wear, and lubrication, this group could focus on how those topics could inform solutions to critical energy problems. Over the next two days, more than thirty experts—professors who led cutting-edge academic laboratories and tribologists who worked for national labs and in industry—met to explore how much energy could be saved by reducing wear and friction. They called this workshop "Can Tribology Save a Quad?" I was a surprise attendee, having had to skip the big annual tribology conference in the preceding couple of years. A testament to the beautiful community that is tribology, an organizer of the workshop saw me at the conference and extended an invitation. I gladly accepted, eager to work collectively with fellow tribologists to tackle such an important topic. I was immediately struck by how international the scientists were, a powerful reminder that tribological solutions and innovations aren't hindered by borders.

The organizers began by explaining that a quad is one quadrillion (10^{15}) British thermal units (BTU), a measure of energy roughly equivalent to the energy produced by 183 million barrels of petroleum, or about 38.5 million tons of coal. The task at hand was to figure out how tribology could save the equivalent of those 183 million barrels of oil annually. We broke out into teams, brainstorming and sharing ideas about tackling issues in transportation, energy generation, and manufacturing. We homed in on what we believed to be the most promising ideas, fleshing them out in finer detail. By the end of the workshop, the participants had surpassed their goal, identifying up to twenty quads of energy savings. To put this in perspective, we currently burn about twenty million barrels of petroleum a day in the United States.[2] A saving of twenty quads is nearly equivalent to half a year's worth of petroleum.

The ensuing report included a call to action with the goal of approaching energy reduction through tribology. The 127-page document identified inefficiencies throughout the entire energy pipeline, in the supply chain, nanoscale technology, and energy auditing, to name a few. After the workshop, the committee shared an online survey with tribologists around the world, asking for their thoughts on how tribology could save more energy. The survey received hundreds of responses from tribologists across the field, in academia and industry. If you've ever had the pleasure of soliciting responses to an online survey, then you know this is an enviable response rate for such a specialized topic. It is a testament to how much passion the tribology community has for the subject of energy conservation.

One of the areas where participants found significant opportunities for energy reduction was, perhaps not surprisingly, the transportation sector. Traditional gas and diesel vehicles are significant producers of carbon dioxide, the most abundant greenhouse gas. The Quad team, as I'll refer to them, identified over two quads of energy that could be saved each year through tribological innovations in vehicles.

One of tribology's greatest success stories has been the automobile. The strides made in improving the efficiency of cars have resulted in remarkable energy savings, while also providing lessons for the design of future vehicles. We're in the midst of transitioning to electric vehicles, but over recent decades, tribological research has focused on the internal combustion engine (ICE) vehicles that still dominate the road. Light-duty vehicles, such as passenger cars, use up more than half the energy consumed by the transportation sector. This includes air, rail, and marine transportation.[3] While we navigate the incredible electrification revolution, the largest disruptor of the industry since Ford began mass-producing cars, we must continue to optimize ICE vehicles. After all, we need to minimize energy waste and emissions today, and ICE vehicles will be on our roads for some time to come.

Over the past decade, two engineers, Kenneth Holmberg and Ali Erdemir, have partnered on seminal studies examining how friction losses affect fuel consumption and emissions in passenger cars. Along with Holmberg's colleague Peter Andersson, in 2011 they revealed what may come as a shocking fact to drivers, that only about 21.5 percent of the fuel we put into our cars is used to move them. Worse yet, approximately one-third of the gas in the tank is used to overcome friction. Even when they removed braking from this figure, 28 percent of fuel was lost to friction. Air drag accounted for just 5 percent of that; friction in the transmission system and in the engine, on the other hand, accounted for about 16 percent of friction losses. These frictional losses amounted to almost 210,000 million liters of fuel in 2009. As you can imagine, these results caught the attention not just of tribologists but of automotive engineers.[4] With over two thousand moving parts in an ICE vehicle, there are ample opportunities to reduce friction and increase energy efficiency.

ICE vehicles have been around for a while, and as is often the case with established technologies, improvements tend to be incremental. Drastic changes don't happen year over year. People have expectations of what their cars should look like and how they should run. Making dramatic changes can backfire. Take the Pontiac Aztek, introduced in 2001, for example. Today this SUV, with its plastic cladding, long, angular body, and disproportionately small wheels, is considered ahead of its time, but twenty years ago, it managed to sell fewer than 120,000 units. It simply looked too different to appeal to buyers. Consumers are especially sensitive to visual changes, particularly changes to a model's body type. We want new cars to look sleeker and more aerodynamic. Such changes don't just appeal to our aesthetics. They decrease drag losses experienced by the vehicle.

Tribologists apply many of the tricks learned from reducing drag on airplanes to designing cars. In place of wings, cars have a cambered airfoil design. If you pull up designs over the decades, you'll notice that most cars have become rounder and shaped more like a teardrop to reduce drag. When oncoming airflow encounters the front of a moving car, the flow is suddenly disrupted, causing pressure to build up. This creates a pressure differential between the front and the back of the car, leading to what is known as flow separation and increased drag. As the higher-pressure air in the front tries to flow toward the lower-pressure air in the back, it creates a drag force opposite the direction of motion. Drag is highest at the rear of the car, where air swirls in vortices due to lower pressure. The elongated back of the teardrop reduces the area containing the vortices behind the car, minimizing the force pushing against the car's forward motion. This is why cars, particularly performance cars designed for speed, are tapered.

The hull design of ships has also influenced cars. The narrower and lower the front of the car, the less drag it experiences because there is less flow separation. However, there's a practical limit to this; people still need to fit inside the car, and consumers want spacious interiors. As with most things friction related, there's a balance of properties to achieve. Other tricks to reduce automotive drag include underside paneling, which limits and streamlines the air that flows into mechanical components, and wheel fender designs that minimize the air that flows into the wheel arches. These designs are first modeled in a computer with air flow simulations that show the effects of drag on a newly designed vehicle. Once engineers are satisfied with their design, the parts are manufactured, and the vehicle is put to the test in a wind tunnel.

Reducing drag by just 5 percent might seem like a small amount. But with millions of vehicles on the road, each driving thousands of miles, 5 percent adds up quickly, both in global energy savings and money you spend filling your gas tank. And, critically, reducing friction, whether drag or another type of friction, has a knock-on

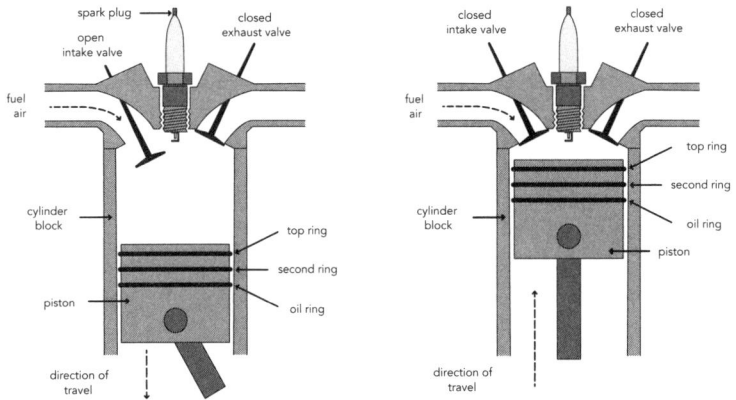

*Piston stroke cycle: intake stage (*far left*); compression stage (*center left*); power stage (*center right*); exhaust stage (*far left*)*

effect: it reduces mechanical power as well as the energy needed for cooling and exhaust.

Other, less visible developments in automotive design cycles are also addressing friction under the hood. Some changes we notice through our bank accounts. When shopping for a car, we pay particular attention to two numbers: the price of the car and its mileage. Improving the latter involves design changes consumers would be happy to see. Still, compared to the average mileage of a car twenty years ago, that number has improved significantly. In 1980 cars in the United States averaged 19.2 miles per gallon. In 2016 they averaged 24.7 miles per gallon. This may not seem impressive until you consider how much more powerful car engines became in this same period. Increased horsepower requires more fuel, but despite the average horsepower doubling during this period, the fuel economy managed to improve.[5] As you may have guessed, this was achieved through the cumulative effects of small changes to our vehicles that reduced friction losses.

The piston assembly of an ICE engine illustrates how tribologists can reduce friction and improve mileage. It converts the fuel you put into the car into the energy to move it. As the piston fills up with fuel,

closed intake valve   closed exhaust valve

cylinder block

top ring

second ring

oil ring

piston

direction of travel

closed intake valve   open exhaust valve

Exhaust

cylinder block

top ring

second ring

oil ring

piston

direction of travel

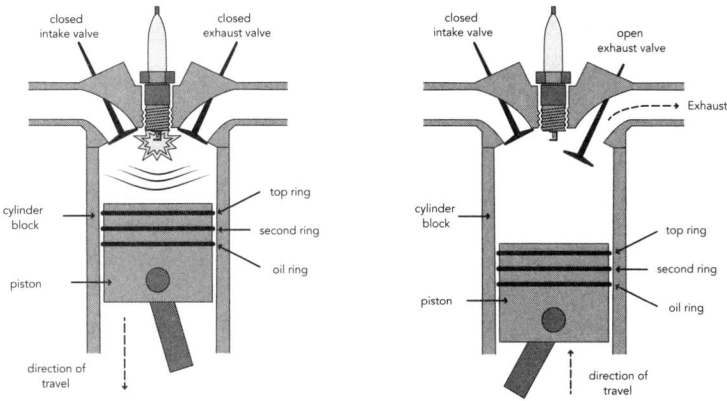

it slides down a cylinder, causing a valve at the top to open. This creates a vacuum inside the piston that sucks fuel and air into the small combustion chamber above the piston. Once the piston reaches the bottom of the cylinder, the valve closes, and the piston begins to travel upward, compressing the fuel and air in the combustion chamber. When the piston reaches the top, where maximum compression is achieved, a spark plug ignites the mixture. The force of the explosion fires the piston back down the cylinder, turning the connected crankshaft, which converts the linear motion of the piston to rotational motion. And voila! Our cars move. In the final stage, the cylinder moves back up from the momentum of the explosion and opens the exhaust valve, pushing out any remaining gases. Cars have multiple pistons, and their movements are carefully synchronized to produce constant motion of our vehicles.

There's a lot of movement going on, so as you can imagine, the piston assembly is one of the most significant contributors to friction in the engine. On average, the piston assembly is responsible for almost half the friction losses in the engine.[6] Within the piston assembly, there are three rings. The top ring, known as the compression ring, acts as a seal, preventing gas and fuel leakage during the combustion process. The bottom ring, closest to the crankshaft, is the oil ring and wipes excess oil from the cylinder wall. Oil rings are designed with gaps, grooves, or bevels to direct the excess oil to the engine's oil reservoir. A thin layer of oil is necessary for the piston to move, but excess oil will burn off and is inefficient. The middle ring, sometimes referred to as the second ring or wiper ring, seals and wipes away excess oil. These three rings alone, with their compression fits and high contact between moving surfaces, can contribute significantly to the friction in the assembly. So can the bottom of the piston, called the skirt, which is tasked with traveling up and down the cylinder to stabilize the piston. A worn skirt can lead to failure to seal the combustion chamber for efficient operation, as well as produce an incredibly annoying noise.

Piston assemblies have undergone numerous changes, although the basic design and function have remained the same. Surface engineering, discussed in Chapter 3, has been key to improving piston assembly operation. As coatings have evolved, fuel efficiency has reaped the benefits. From simple MoS_2 sprayed coatings to highly engineered PVD coatings of DLC, surface engineering has reduced friction and increased wear resistance, improving the longevity of parts and reducing waste. When parts last longer, less energy goes into producing and replacing them.

Surface texturing has also improved the efficiency of piston assemblies, but it comes with its own set of challenges. With seemingly infinite possibilities for surface texturing patterns, there's no practical way to experimentally try them all. Physics can get us started, providing optimal combinations of shapes, sizes, and depth, but it is time-consuming to physically produce and test each combination. Even if you somehow knew with certainty that a specific surface pattern design would be successful, slight variations in depth from the manufacturing process could alter the successful execution of it. Variations as small as nanometers can change how a lubricant flows, impacting overall friction. Manufacturing will be an expensive and time-consuming endeavor, at least until technology is implemented to enable high-precision, high-throughput texturing.

Some tribologists model the problem computationally, working out how to program the fluid mechanics of various lubricants under real-world engine conditions for temperature and speed. They build a system in their computer models and run simulations, employing various surface texturing patterns—dimples to chevrons and everything in between—to predict how the lubricant will flow and determine the corresponding friction behavior. The most promising patterns can then be fabricated in the lab with prototypes made for bench top testing. Lab scale testing for automotive applications runs the true gamut of tribological testing. Some setups take up merely a couple of square feet on a bench top; others take up dozens of square feet, spanning entire labs. The data from these tests can be fed back into the model to refine it and improve its accuracy in predicting lubrication and friction. After this proof-of-concept stage, the winning design (or designs) is tested on a full-scale car engine. Finally, the design with the optimal surface texture makes its way to vehicle testing, where it is implemented in an engine by a manufacturer and run through the standard series of automotive qualification tests. If it satisfies their performance requirements, consumers reap the fuel saving rewards.

During my postdoctoral work, I visited a lab at Loughborough University, where I was fortunate to witness a multidisciplinary team of automotive engineers and researchers from multiple universities come together to tackle friction in the piston assembly. One group of researchers designed simulations of the assembly, including any design modifications the team determined would be advantageous. Another group prototyped surface texturing patterns. And a third group developed experimental methods to measure properties like friction and oil film thickness. Their objective was to collect real-time information on the performance of the three rings in the assembly. Using ultrasonic sensors, discussed in Chapter 4, they monitored the lubricant thickness during the four stages of the piston cycle as it moved up and down the cylinder. With such information, the third group could test and validate the work of the other two groups.

This work began by testing the proof of concept on a lawn mower engine to determine the ideal placement of the ultrasonic sensors.

The results were then used to instrument a full-scale engine test bed. With the data from the ultrasonic sensors, the researchers could answer such questions as: Is the lubricant present at all times, and what lubrication regime is it in? Does the texturing help push a boundary lubrication condition to the more desired hydrodynamic condition? Or does it do the opposite and starve the contacts of lubrication?

The data provided insight into the lubricant itself, enabling the tribologists to develop various formulations that could be tested to find the optimal one.[7] In addition, the team could determine whether the surface texturing patterns were having a positive or negative impact on lubrication and friction. If the film thickness remained constant over repeated cycles and then the thickness suddenly changed, that was a sign the ring was wearing. It sometimes takes large project teams to tackle what seems to be a small problem: optimizing piston assemblies. But it is by tackling these niche friction problems that we achieve larger goals, such as saving a quad of energy and reducing greenhouse gas emissions.

Cutting Emissions Through Friction

Fuel efficiency, and thus energy savings, aren't the only challenges sending ICE vehicles to the end of their motorway reign. In the 1990s, and more recently, in 2015, international collaborations such as the Kyoto Protocol and Paris Agreement, which aimed to address climate change globally, brought CO_2 emissions to the forefront of news headlines and drivers' minds. A decade ago, nearly 20 percent of anthropogenic greenhouse gas emissions stemmed from the transportation sector, with 80 percent of that coming from the roads. But in 2009, the European Union (EU) imposed emissions standards on light-duty vehicles, with the intention of reducing greenhouse gases 20 percent by 2020. In 2019, Holmberg and Erdemir followed up on their initial work on fuel efficiency, showing that by implementing tribological solutions, such as the application of surface texturing in piston assemblies, it was possible to effect significant reduction in carbon emissions. They estimated that over 290 million tonnes of CO_2 could be saved worldwide over five to ten years, just shy of 1 percent of total annual global CO_2 emissions. Projecting out further, they estimated that in twenty-five years, it would be possible to reduce CO_2 emissions by 960 million tonnes, over 2.5 percent of current annual emissions.[8]

Sometimes, the engineering solutions to reduce emissions make tribologists groan because of the tribological problems they may cause. Take, for example, the stop-start engine. Over half of new cars sold in the United States are equipped with automatic stop-start systems. And yet, somehow, they still catch me off guard when I'm walking around my neighborhood. I don't expect a car to shut off and turn on just to allow me to use the crosswalk. The technology has been around since the 1970s, when Toyota first introduced it, but it didn't become widespread until the 2000s, when emissions regulations were introduced. A car that idles for ten seconds or more releases more emissions than a car starting. That is why automatic stop-starts play such a key role in reducing CO_2 emissions.[9] Tribologists are fully on board with the reduction in emissions, of course; the problem is that it throws a wrench into the well-designed lubricating systems the tribology community has spent decades optimizing.

If you recall from the fluids chapter, there's a sweet spot on the Stribeck curve where solid surfaces are separated by a thin lubricating film. Achieving a hydrodynamic lubrication regime is the goal in automatic stop-start engines, but turning the car off and on again is not ideal for achieving it. While the car is running, the engine systems provide oil flow to achieve and maintain hydrodynamic lubrication of parts like the piston assembly. Cutting off that flow leaves the assembly in boundary lubrication. There might be drops of oil between some asperities, but for the most part, there is solid-on-solid contact. As a result, the startup, or static, friction is high. Then there's perhaps the bigger problem, which is wear. The more rubbing between solid materials, the higher the wear. A vehicle without automatic stop-start technology typically experiences about 50,000 stop-start events; compare this with up to 500,000 events for a vehicle with automatic stop technology.[10] Higher wear is one of the biggest concerns with automatic stop-starts. The last thing anyone wants is a solution that is good for emissions but bad for material waste and production because drivers have to replace parts more often. Emissions, after all, happen during part production and delivery as well.

As you've likely guessed, tribologists are on the case. If lubricant circulation is needed when the engine cuts off and the brakes are fully engaged, the obvious solution is to find a way to continue circulating it. Most stop-start cars have an electric pump that provides continuous fluid flow, maintaining hydrodynamic lubrication while the vehicle has stopped. This pump is independent of the engine of the car, enabling it to circulate lubricant even when the vehicle stops moving and the engine shuts down. That's an engineering design solution, as opposed to a material formulation solution. In cases where this can't be implemented due to space or other design constraints, tribologists turn to surface engineering, matching the correct coating and oil to the engine. And sometimes, they combine both solutions.

With more wear events occurring, having a durable coating is imperative. So too is ensuring that friction remains low. Otherwise, the adhesion between parts can be so high upon restarting the engine that nothing will move. In recent years, researchers have been experimenting with the use of nanoparticles that act as minuscule bearings in engine oils. Under high pressure, which asperities experience upon stopping, the particles compress into nanoflakes, sticking to the solid surfaces and providing a lubricant film. This way, when the engine starts up again, the friction can be low, despite the fact that the oil delivering these nanoparticles has since trickled out of the contact.[11] To some, idea seems too good to be true. They are skeptical about the efficacy of nanoparticles, especially over repeated use. Others have concerns about the potential health and safety risks associated with such small particles that we may inhale.

One variation of this solution involves oxidation. Rust may seem counterintuitive to any automotive problem. After all, engine lubricants are formulated specifically to prevent corrosion. But some coatings, called ferrofluids, are mixed with rust nanoparticles, or iron oxide, to help keep friction and wear low.[12] Iron oxide is ferrimagnetic, meaning it exhibits permanent magnetism because of the intrinsic spin of the electrons in the iron ions. When the car starts up, ferrofluids flow quickly, guided by the magnetic field generated by the engine. Researchers in Pakistan are even looking into a more environmentally friendly version of engine lubrication, using castor beans to produce the base oil instead of oil made from petroleum.[13] Castor oil has been used as a lubricant since World War I. Castor beans thrive in tropical and subtropical climates, enabling them to be cultivated across the world. The researchers found that adding iron oxide to this biolubricant decreased its friction by about 50 percent and reduced wear by 40 percent.[14]

This is all well and good, but with ICE cars set to become a thing of the past, what do all these improvements mean for the future? Fortunately, these tribological innovations can be used in hybrid vehicles and vehicles that rely on alternative energy sources. Perhaps most importantly, they can be used in electric vehicles (EVs), even though the number of moving parts and types of parts are drastically different than in ICE vehicles. The piston assembly, where tribologists have achieved large friction reductions through a combination of texturing, coating, design, and improved lubricants, doesn't even exist in EVs. However, surface texturing can be applied to other moving components in the EV. The knowledge and experience gained through understanding texturing has continued to be fruitful for EVs.

The design of the EV itself naturally leads to lower friction: while the engine of an ICE vehicle can have thousands of parts, an EV engine has fewer than a hundred. With so few parts, the frictional losses in EVs are much lower. In fact, the energy usage in the two vehicles is almost completely inverted. EVs and ICE vehicles have the same power requirements. But whereas an ICE vehicle tends to use only about a third of its fuel to move, with the rest essentially wasted through energy losses from heat and friction, an EV typically only *loses* about a third of its energy. Around 75 percent of the energy put into an EV goes directly into moving the car, and the remaining 25 percent is again used to overcome various energy losses.[15]

Breaking down these numbers for EVs is more complicated owing to features like regenerative braking, whereby energy that is typically lost during braking is instead used to recharge the battery. With regenerative braking, the motors that enable motion also stop it, turning in reverse to slow the car down. As the motors turn backward, they act as generators, creating electricity to charge the vehicle's battery. Traditional friction-enabled braking is used as a backup to this regenerative braking system.

There are different ways to think about the energy consumption and waste of an EV. Are we strictly concerned with how much energy taken from the grid is converted into motion, or do we want to consider the energy recovered through regenerative braking? If we're just summing up energy usage, an EV actually consumes more than 100 percent of what it takes from the grid because of how regenerative braking works. Like an ICE vehicle, moving an EV requires energy to overcome rolling friction and drag. But unlike an ICE vehicle, braking in an EV converts the kinetic energy into electrochemical energy. This not only reduces the friction losses caused by braking but also recovers another 6 percent of energy from braking. This means the EVs are actually consuming 106 percent energy.[16] I find it incredible that my car can consume more energy than what I initially put into it. Of course, friction still causes some thermal energy loss during this braking process, but far less than it does in an ICE vehicle. Assuming that tire friction and drag losses between these vehicles are comparable, we're looking at friction losses of 16 percent for ICE vehicles versus about 6 percent for EVs. This is clearly significant progress in energy conservation with regard to friction reduction, but that doesn't mean friction is no longer a concern. There may be less of it in the overall picture, but it's still there, and energy loss is energy loss. As we saw with the impact of incremental frictional improvements in the efficiency of ICE vehicles, there is ample opportunity to keep moving that needle.

The hard work of automotive tribologists is paying off with EVs. Even without a piston, many of the techniques developed for reducing friction in ICE vehicles—surface texturing, reducing the viscosity of lubricants where needed, and low-friction, low-wear coatings—come in handy. But sometimes, tribologists have to get creative when testing such solutions. With the electric motor comes an increase in copper parts in the vehicle, and these copper parts are susceptible to corrosion. Lubricants can help prevent corrosion, but it's imperative that the engine oils be appropriately formulated because of vapor phase lubrication. As the engine heats up, volatile components in the base oil can vaporize and be corrosive when they land on and react with copper. These vapors can be insidious, creeping into areas where liquid lubricant isn't present to offer protection. To combat this, researchers have developed new ways of testing EV lubricants, using electrical resistivity measurements. As copper corrodes, the corrosive layer impedes the flow of electrons in the material, increasing its resistivity. Running a current through a copper wire and measuring its resistivity offers real-time insight into potential corrosion. To do this, tribologists submerge a copper coil in a lubricant bath and place another copper coil outside the bath in the path of potential vapors. The apparatus is set to the temperatures that the lubricant will experience in the EV, and then a current is run through the coils. If the measured resistance of the dry coil remains stable and comparable to that of the submerged coil, tribologists know they have a successful formulation. On the other hand, if the resistance of the dry coil decreases, then the lubricant has corroded the copper.[17]

The elephant in the room, whether a vehicle has an electric motor or piston, is the tire. Tires can cause significant energy loss through rolling friction. Tires need some friction for traction, but rolling resistance also tries to prevent the wheels from rolling at all. To maintain a constant speed on the road, the tires must overcome this resistance. For EVs, this can be a nightmare because rolling friction is linearly dependent on the mass of the car. Unfortunately, the batteries used in EVs add a significant amount of weight to a vehicle. Sure, the engine might be lighter, but you're still potentially looking at over a hundred kilograms of added weight from the batteries. One way to reduce rolling resistance is to keep finding ways to lighten a vehicle, and in particular, battery packs.

Teams of researchers are working on this problem, developing lighter composite materials and smaller, more efficient batteries. All eyes are currently on solid-state EV battery development. Lithium-ion batteries work by flowing lithium ions from the positive side of the battery, called the cathode, to the negative side, called the anode, or vice versa, depending on whether one is charging or discharging the battery. The efficiency of the battery depends on how well an electrolyte mixture transfers the lithium ions between the charged ends in all driving conditions. Traditional lithium-ion batteries use a liquid solution made of a solvent and lithium salt, which can swell and be thermally sensitive, making it prone to explosions. These batteries are also less efficient at lower temperatures, leading to complaints about how inefficient EVs are in the winter. On the other hand, solid-state electrolytes offer higher energy per unit volume since they're more densely packed, meaning they can save space and, crucially, weight. Solid-state batteries hold up to three times more energy than their liquid counterparts and are less volatile, making them a safer option. Currently, solid-state batteries are used in small devices, like my watch. The primary challenge has been to scale them for more energy-demanding EVs. If successful, solid-state batteries would be a game changer for the EV market. Batteries would last longer and be safer, and packs would be a third of the size. Cars would become more efficient, and from our tribological perspective, the reduced weight would help reduce rolling friction.

Another way to reduce rolling resistance is through tread compound formulation to minimize hysteresis. *Hysteresis* is the lag time of a materials' physical response, such as deformation, to a force. As one loads and unloads rubber materials, the molecular chains stretch and recover. But in that process, the chains respond at different recovery speeds, causing them to rub against each other and create frictional heating. The heating softens the material so that more chains come into contact with the road, increasing traction. But, since friction converts energy to heat, hysteresis results in a loss of energy.

Unfortunately, since hysteresis in the treads is also responsible for traction, finding the right tire material can be a tricky balancing act.

Hard rubber compounds, for example, have less hysteresis and will recover their original form, but this means less material deforms and sticks to the road to provide traction. Softer tires, on the other hand, grip the road more but result in higher fuel consumption to overcome the energy loss.

Fillers in tire compounds have helped to solve this problem. Beginning in the 1990s, automotive engineers began adding silica (silicon dioxide) filler to tire rubber. Most silica must be mined, so it has an environmental cost.[18] Recently, efforts have focused on making more sustainable tires by using bio-derived silica. In 2020, a team of scientists at Tata Chemicals and Savitribai Phule Pune University in India not only derived silica from rice husk ash but demonstrated its effectiveness as a tire filler. Rice husks are a significant by-product of milling rice, accounting for almost 25 percent of the total weight that is harvested. These husks are used in boilers to generate energy, leaving behind rice husk ash that accumulates and can be an environmental hazard.[19] Around the world, researchers continue to develop new materials for improved tire performance, seeking to optimize the balance between rolling friction, traction, and hysteresis. Such materials could offer around 3 percent in fuel savings, and sometimes even more, depending on the vehicle and driving conditions.[20]

The Sound of Friction

As we make the transition from ICE vehicles to EVs, there's another reason that friction in EVs shouldn't be written off as insignificant. One of the perks of EVs is that they're delightfully quiet. Until they aren't. Without a combustion process, small noises that we would otherwise never notice suddenly start chirping away in the background. It's like the whine of my computer monitor; when I have a fan running in the room, I don't notice the monitor's high-pitched squeal. But when I flip off the fan, I can't unhear it.

Friction can cause vibrations, such as when stick-slip occurs, and these vibrations can reach audible frequencies. In ICE vehicles, the noise of parts like vibrating gears is drowned out by the engine cycle itself. EVs, while more efficient, still have plenty of moving parts that can lead to friction noise. Some of the most significant culprits of noise in ICE vehicles—shock absorbers, brakes, and windshield wipers—seem even louder against the relative quietness of the car.[21]

Windshield wipers are quite the fun tribology problem. Designing parts for low friction in constantly changing conditions—in this case, the amount of water present—is no easy feat. You may have noticed that sometimes your wipers work perfectly. Then they start chattering along, either because you had to change their intermittent speed or because the intensity of the rain has changed. This instability is connected to the Stribeck curve and different lubrication conditions. In particular, the wipers are prone to squeaking and chattering between the boundary and elastohydrodynamic (EHL) lubrication regimes. When this happens, feel free to shout encouragement at the rain conditions to ease them into EHL; it's more productive than road rage.

Designing windshield wipers is challenging for multiple reasons—and not just because the slightest change in raindrops may alter the lubrication regime. Windshield wipers need to be stiff enough to clear the rain and snow but not so stiff that they scratch your windshield and squeak. Tribologists incorporate various rubber compounds and additives, as well as surface engineering, into their designs. Using contact models, such as Greenwood-Williamson's or Persson's, they can design the geometry and surface finish of the wipers to achieve the desired stiffness and contact. They then simulate various amounts of stress and deformation, as well as different fluid dynamics.

Despite the challenges of noise from friction, EVs are crucial in our fight against climate change since they don't produce any direct emissions. ICE vehicles can produce up to 4.5 times the amount of CO_2 produced by an EV. The key word being: can. This number assumes that the electricity powering up the EV comes from renewable sources. If the electricity used to charge the car comes from coal, EVs, disappointingly, produce more emissions than an ICE vehicle: 228 g/km versus 224 g/km in ICE vehicles, which includes fuel production and manufacturing. Oil-generated electricity brings the EV CO_2 emissions slightly below that of ICE emissions, and natural gas reduces emissions by nearly 100 g/km less CO_2, a definite improvement but still more than double what can be achieved with electricity sourced from renewables. It's a sobering reminder of why we must continue to push for renewable energy sources such as solar, hydropower, nuclear, wind, or geothermal.[22] It's especially sobering when we consider the reality that manufacturing EVs produces more CO_2 than ICE vehicles. Additionally, it turns out that batteries are not environmentally friendly to produce. Mining the materials used in batteries—including lithium, nickel, and cobalt—has a high environmental cost, producing toxic fumes. The extraction process is also water intensive. Water depletion in mining zones has left large areas of land devoid of wildlife. As we transition to more sustainable transportation, it's imperative that we all keep an eye on the larger picture—the entire life cycle of the car.

Suck, Squeeze, Bang, Blow, and Friction

At the peak of the 2020 Covid-19 pandemic shutdown, people marveled at how quickly the air quality improved. The lack of road travel seemed to have an almost immediate impact. So too did the unprecedented grounding of flights, as airlines found themselves scrambling to find parking for planes that used to shuttle back and forth on busy routes. Road travel might make up the majority of energy consumption in the transportation sector, but the nearly 10 percent that aviation consumes is nothing to bat an eye at. In the past decades, much progress has been made. Planes are 70 percent more efficient than they were just forty years ago with the introduction of revolutionary composite materials and innovations like winglets, discussed in Chapter 4.[23]

Likewise, innovations in engine design have significantly improved aircraft efficiency. The concept behind a turbine engine is, surprisingly, much simpler than that of a car engine. When I was an undergraduate, I interned with a major aircraft engine manufacturer. On the first day of the internship, during onboarding, I was told to memorize four words, in this order: "suck, squeeze, bang, blow." If your reaction to that was "pardon me?," then you can relate to my confusion. These four words are a succinct mnemonic for remembering how a jet engine works. The fan sucks in surrounding air; the compressor squeezes it, causing pressure to build up; the high-pressure air combines with fuel in the combustion chamber, where it ignites; and the hot gas is expelled to a turbine to propel the plane forward. It's a beautifully simple concept with incredibly complex, highly engineered parts making it happen in the most efficient way possible.

The only part of these engines that we see are the fan blades of the fan assembly. But behind that assembly, hidden from our view, are thousands of small moving parts in the compressor assembly. The hollow cylindrical compressor is lined with rows of airfoil-shaped vanes, tasked with increasing the pressure of the air flow to optimize conditions for combustion. There are actually two compressors in the engine—the low-pressure compressor and the high-pressure compressor—connected by a shaft running through the center of the engine. Immediately behind the fan is the low-pressure compressor. It starts pressurizing the inlet air for the high-pressure compressor. This compressor squeezes the flow, creating so much pressure that the air molecules want to expand to release the pressure. This forces the air into the combustion chamber, where it mixes with fuel and is ignited. The more pressurized the air, the more efficient the combustion process.

Compressors contain stages, each of which pressurizes the air more than the previous stage. The number of stages depends on the engine; for example, the newer, more efficient engines used for transatlantic flights have ten stages, whereas their predecessor had fourteen. Each stage consists of a row of blades, known as rotor blades. The blades rotate perpendicular to the shaft, directing air through the engine. Between each row of rotor blades is a row of nonrotating vanes, called the *stator,* which increases the pressure of the flow. If you looked along the length of a compressor, the rows of rotor and stator vanes would almost look like zigzags. As the air leaves the rotor blades and hits the stator vane, the flow velocity decreases. The kinetic energy is converted to internal energy, resulting in an increase in pressure. More pressurized air then flows into the next rotor, repeating the process. These vanes are contoured through fluid dynamic design, with their airfoil geometries sometimes varying from engine to engine to further optimize airflow. At the end of the compressor stages, the pressurized air is fed into the combustion chamber.

If the speed of the airflow becomes too high, the airflow will no longer be smooth and unidirectional but will instead break down, bouncing off the vanes. This ultimately results in less efficient pressurization. In extreme cases, this can cause the compressor to stall, leading to catastrophic outcomes. More common in early jet engines, compressor stalls have been addressed in modern engines by installing variable stator vanes (VSVs), typically in the first few stages of the high-pressure compressor. These vanes are attached to the stator, which does not revolve about the shaft. Traditionally, the rows of stator vanes were stationary, but VSVs are able to rotate in place, where they are attached to the stator. VSVs redirect airflow so that the air hits the rotor blades at an angle the blades can tolerate, preventing the flow from breaking down and ensuring efficient engine operation. They do this by adjusting their angle. During takeoff and landing, the vanes rotate up to thirty degrees around their individual shafts, mounted in the stator.[24] During flight, the vanes continue to rotate as conditions change, but usually no more than a couple of degrees. It's hard to believe such small movements can matter. With the speeds and pressures of a jet engine, being able to adjust the inlet and outlet areas of the airflow ever so slightly affects pressurization enough to impact fuel efficiency. This is why modern engines are able to fly with fewer stages than the previous generation of engines.

These subtle movements are possible thanks to an innovation called FADEC. FADEC, or Full Authority Digital Engine Control, is a computer-generated engine control system that optimizes the engine parameters to, literally, get more bang for the buck while also providing a quieter engine. Its evolution over the decades has been aided by advances in computational technology. FADEC takes flight parameters such as air density, engine air pressures and temperatures, and throttle position, which controls the amount of air and fuel flowing into the engine. It then calculates operating parameters like air flow, fuel flow, and ignition timing. FADEC transformed aviation. Electrical controls replaced mechanical levers, resulting in lighter engines and improved safety. Whereas pilots once had to manually contend with valves and rods to control increasingly complex engines, FADEC automatically adjusts ignition and the mixing of fuel and air. FADEC instructs the variable stator vanes to switch on, and then it finds the position for the vanes that achieves the most efficient air flow. The vanes need to respond swiftly and not require so much energy that they become counterproductive. Spoiler alert: friction will now be entering the discussion.

With VSVs constantly moving due to FADEC, friction and wear become a concern. Rather than mounting the vanes in the stator casing, causing metal on metal or metal on composite contact, they're housed in good old-fashioned bearings. Shaped like little top hats that are missing their crowns, these bearings, called *bushings,* form a relatively large amount of apparent contact area with the shaft running through them. Often, they're made from high-performance polymers with solid lubricants to provide a combination of low friction and low wear. Static friction must be low enough that the actuator, which converts fuel into kinetic energy to provide motion to the VSVs, can overcome it easily. Otherwise, the vanes won't be able to continuously adjust in flight. In addition, the bushing must be light enough that it doesn't significantly add to the engine's bulk, as weight reduces fuel efficiency. This has a knock-on effect. The lower the friction between the bushings and the vanes, the smaller the actuator needed to move them and the more efficient the engine is.

Another way to increase the efficiency of the engine is to run it hotter. The hotter the combustion chamber, the less energy is required for ignition. That leaves more energy available for additional thrust, the force that propels the plane forward. In other words, more mileage for the same amount of fuel. Hotter engines also mean that materials must withstand higher and higher temperatures. Since jet engine temperatures can reach over 1000 °C, polymers and traditional lubricants can't be used. The material options are metals and ceramics. To counter the additional weight of these materials, composites must be designed to combine temperature capability with a lighter-weight package. Tribologists are also hard at work pushing the temperature limits of low-friction coatings to meet these needs. Bushings still need to be made from a low-friction material so that the coating doesn't wear away. One solution has been ceramic matrix composites. Fibers such as carbon, woven into a ceramic matrix, offer lighter weight and good strength, thermal, friction, and wear performance.

Sometimes, increasing efficiency requires engine redesign. In recent years, possibly the most controversial innovation in aircraft engineering was introduced to the jet engine: geared turbofans. This engine consists of a gearbox that enables the fan and turbine to spin at different speeds, requiring fewer compressor stages, reducing the weight of the engine, and increasing its efficiency. The concept, proposed in the 1980s by engineers at the aerospace company Pratt and Whitney, was met with criticism. Skeptics contended that the high loads and speeds required of the engine parts would in turn require a large, bulky gearbox. Such a large gearbox would ultimately add more weight than would be saved by having fewer compressor stages, making it less efficient than a traditional engine. There was also the question of maintenance. Some speculated that gearboxes would need more frequent maintenance than the typical jet engine to ensure the gears were not wearing.

Pratt and Whitney refused to fund the endeavor. Not wanting to abandon the concept, small teams of engineers worked on the engine off the books, out of the senior leadership's line of sight, until the early 1990s, when a demonstration engine was finally completed.[25] The company then needed to scale the proof of concept to a production engine to prove the skeptics wrong. Work began in earnest in the late 1990s, but as deadlines and budgets slipped, the project raised further eyebrows. Was Pratt and Whitney crazy for staking so much on what most in the aviation industry considered a hopeless gamble? At tribology conferences, engineers used to snicker at the idea under their breath, doubting it would ever come to fruition after decades of effort and over a billion dollars spent. But in 2008 the aerospace company finally brought the geared turbofan to market.

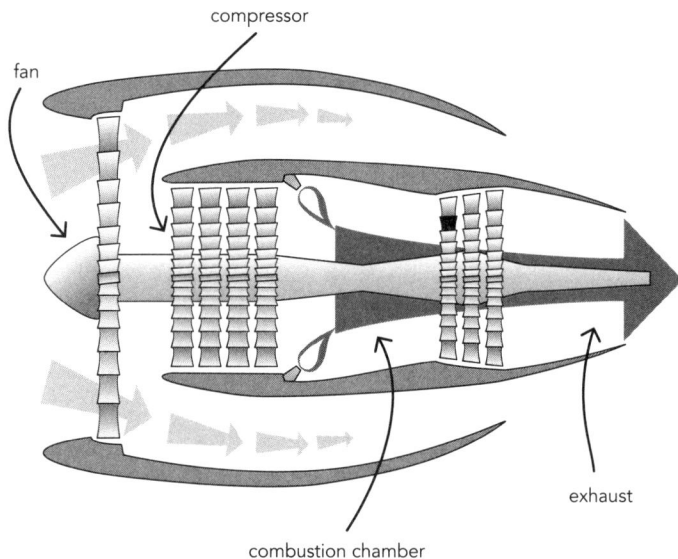

Traditional turbofan aircraft engine (left); geared turbofan aircraft engine (right)

Geared turbofans have dramatically reduced noise and increased efficiency. A typical jet engine has one shaft connecting the low-pressure components, such as the fan and compressor, to the low-speed turbine. A second, concentric shaft connects the high-pressure components. The fixed shaft means the rotational speed of the components connected to the fan is limited by the speed of the fan blades. A geared turbofan instead contains a gearbox between the fan and compressor, enabling the two parts to be decoupled. The gears inside the gearbox rotate at different speeds, allowing the fan to run at lower speeds and the low-pressure compressor to rotate faster. Because the compressor can now rotate faster, fewer stages are needed to pressurize the air. This design results in weight and fuel savings and a more efficient engine overall. And with a lower fan speed, the engine itself is significantly quieter.

For all this to work, friction in the gearbox must be mitigated. Because the gearbox is located behind the fan, in the cooler part of the

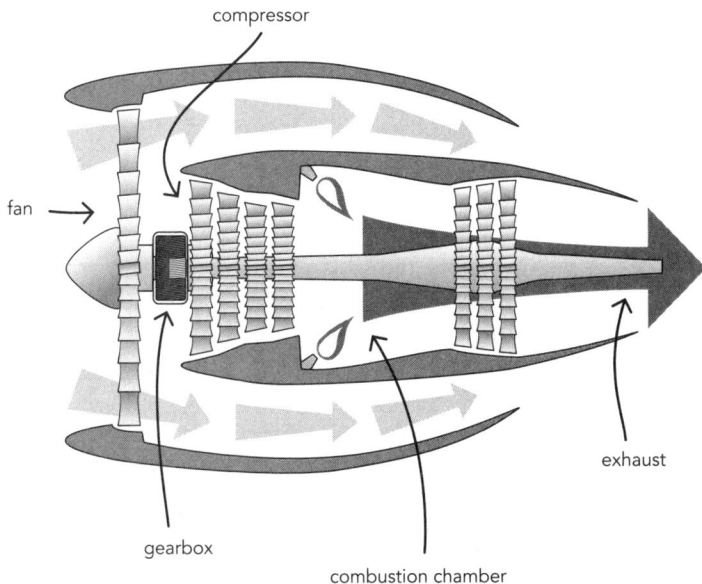

compressor

fan →

gearbox

combustion chamber

exhaust

engine, oil-based lubricants can be used. The synthetic base oils used in aviation have good thermal stability. But they are more expensive due to the synthesis process needed to make their complex molecular structure. One of the jobs of the aviation lubricant as it circulates is to transfer heat away from a surface. This is particularly important in high-temperature jet engines. Unfortunately, many thermally conductive additives that may enhance heat flow produce ash or sediment that can destroy the engine. Tribologists continue to seek ways to develop lubricants that reduce thermal heating using complex carbon-based chemistries that are safe to use in engines. Lubricants in the gearboxes in turbofans must also be able to withstand the high loads experienced by the gears.[26] If they can't handle the load, they will break down, leading to solid-on-solid contact, high friction, and high heat, which can cause them to ignite. Formulators are working on enhanced versions of aviation lubricants to meet this need and reduce the maintenance intervals required to ensure gearbox performance. In 2016, the first airliners equipped with the storied geared turbofan took flight. According to data from Pratt and Whitney, the geared turbofan not only has thrust on par with any other traditional jet engine but is 15 percent more fuel efficient and up to 75 percent quieter.[27]

Most disruptive technology fails to roll out without a hitch, and the geared turbofan has had its share of hiccups. The list of issues that needed further attention looks like the topics of a typical tribology conference: seals, corrosion, and lubrication. Inside the gearbox is a high-speed planetary gear system that cannot be impeded by friction. Lubricating oil keeps the friction between the gears low and prevents corrosion, which can lead to premature wear and, ultimately, part failure. But during testing, contaminants from these oils leaked into the compressor, requiring engineers to return to the drawing board and redesign the gearbox assembly.

Today, around one thousand of nearly 26,000 commercial aircraft sport the geared turbofan. Over the coming years, we'll get a better picture of the performance of geared turbofans, and whether they can live up to their marketing hype. In fairness, other new engine technologies that don't involve such drastically new designs have faced their own issues upon introduction. There's also the other type of friction—social friction—that surrounds this engine. If you happen to meet someone who works in the aviation industry, ask them about their thoughts on geared turbofans. I assure you it's a lively conversation starter.

We've focused on commercial aviation, but that's not to suggest military engine development is stagnant. Far from it. The performance requirements for commercial and military engines differ significantly. Commercial engines are optimized for fuel efficiency and noise reduction. If you've ever heard certain types of military jets fly over, you know that noise reduction is clearly not a top priority when designing those engines. Their job is to fly fast and be highly maneuverable. Some can travel faster than the speed of sound, resulting in sonic booms that scare the daylights out of you if you don't know one is training nearby. To achieve extra thrust, they're fitted with afterburners, tails in the back that glow red and orange from the flames they produce.

The aim of the game in military engines is to meet performance requirements first, then worry about fuel efficiency. But let's also be honest here when it comes to the motivation behind fuel efficiency. In most countries, the military looks at fuel efficiency in terms of the ability to fly longer missions, not necessarily low carbon emissions. Fuel efficiency also saves money. In the United States, jet fuel is approximately 80 percent of the Air Force's energy budget.[28] But whatever the motivation to become more fuel efficient, we'll take it. Energy saving is energy saving, and we need every bit of it to combat climate change.

Governments have begun stepping in, pushing for caps or, at the very least, transparency around military emissions. Virtual flight simulators are one way to reduce emissions and save energy. Military training can often be carried out without burning the fuel of an actual aircraft. However, a proper solution needs to be more holistic, addressing the physical problem itself, as well as finding workarounds like simulators. This brings us back to engine and aircraft design.

One promising solution has been a variable or adaptive cycle engine that is designed to handle a variety of flight conditions experienced by military aircraft. A typical engine is designed for optimum efficiency under specific flight conditions—either subsonic, transonic, or supersonic speeds. The high-speed Concordes, for example, were terribly inefficient and loud at takeoff and landing but happily cruised at supersonic speeds. Variable cycle engines maintain efficiency as conditions change by adjusting airflow, switching between the high thrust of a typical fighter jet and the high efficiency of a typical commercial airliner. It's one of those innovations that seemed like science fiction, seemingly out of grasp. Not anymore. In 2021, GE completed testing of its adaptive engine, which is not only 25 percent more energy efficient than a typical fighter jet but offers improved thrust as well. The adaptive engine consists of a three-stage adaptive fan that provides an extra stream of air compared to a typical engine. This airstream can be used for extra power or diverted elsewhere in the engine for cooling. Doing so enables the adaptive cycle engine to alternate between high-efficiency mode, in which the engine temperature can be regulated by the cooler air, and high-thrust mode, in which the extra airstream is used to achieve more thrust.

The three-stage design also helps address another form of friction. Spillage drag occurs when the inlet air is more than the engine can swallow and "spills" around the outside of the engine instead of flowing through the compressor. Typically, spillage drag happens when the pilot throttles down or decreases power. The engine is confronted with the same amount of air due to its inlet size, but because it needs less power, it draws in less air. As the residual air flows around the engine, it increases drag and reduces efficiency. Spillage drag is more of a concern for military aircraft, which travel at higher speeds. The adaptive engine controls the airflow through its third stage, allowing the engine to continue to ingest the same amount of air and directing it for thrust or cooling, even as throttling is decreased. This ensures that airflow remains constant, even in lower power modes.[29]

The adaptive engine was initially selected for the Lockheed Martin F-35A aircraft family and expected to enter service around 2027, but it failed to win a US Department of Defense contract over concerns about the costs of upgrading an entire fleet with the new technology. Instead, the Department of Defense chose a competing, more traditional engine from Pratt and Whitney. Adaptive engines have spawned the latest showdown between behemoths Pratt and Whitney and General Electric (GE). With GE's test run completed, Pratt and Whitney is racing to complete its competing adaptive engine, continuing a rivalry for US defense contracts, a rivalry that dates back to the 1980s, when GE broke the Pratt and Whitney monopoly with the US armed forces by landing a contract to put their engines on the F-16 fighter jet. Now the race is on for the next generation of military engines, with all eyes on these new adaptive engines.

Reducing Energy to Produce Energy

As mentioned earlier, the transportation sector is responsible for around 30 percent of global energy consumption. This is significant, but it also means that there's another 70 percent of energy usage to address in other industries. Friction accounts for one-fifth of *all* energy consumption.[30] That amounts to over 100 exajoules (EJ). For comparison, the 9.0 magnitude earthquake that devastated the Tohoku region of Japan in 2011 released roughly 0.2 EJ of energy.[31]

Manufacturing is one of the top four energy-consuming sectors globally. About 20 percent of the energy consumed in manufacturing is used to overcome friction.[32] The bottom line is that manufacturing generates considerable friction, especially subtractive manufacturing processes in which parts are made by removing material through cutting, milling, or grinding. There's friction between the cutting tool and the material it's shaping and between the cutting tool and the spindle turning it. That spindle is powered by a motor whose moving parts create plenty of friction. In news that would make Thurston cringe, from a quarter to half of the mechanical work put into a cutting tool is associated with friction between the cutting tool and workpiece.[33] As far as friction is concerned, this interface is an incredibly severe environment. A freshly cut material has no oxide formation on the surface to help reduce friction. A metal part can cause high metal-on-metal adhesion, resulting in detachment of material from the cutting tool, especially at high temperatures that are easily reached in machining parts.

If you visit a machine shop, you will see plenty of metal-cutting fluid flowing directly onto the cutting tool during operation to prevent this adhesion. At lower cutting speeds, these fluids lubricate the contact zone. However, at higher speeds, it's easier for the fast-moving cutting tool to spray the lubricant in various directions. Between the tight contact of the cutting tool with the workpiece and the rapid movement of the cutting tool, it's difficult for the fluid to penetrate the contact between the tool and material being cut to reduce friction. This yields the potential for high adhesion, leading to high wear. To reduce friction and the cutting temperature, many tools are coated with a titanium alloy to improve its thermal and tribological properties. Very thin coatings of these alloys, as well as other antifriction and antiwear coatings such as DLC and MoS_2, can be applied via physical or chemical vapor deposition, discussed in Chapter 3. While costing more up front, these coatings ensure longer-lasting tools while significantly reducing the friction of moving parts and energy requirements.

Of course, a significant way to reduce energy consumption is through innovations in the energy industry itself. It's estimated that energy losses due to friction in the energy industry could be as high as 20 percent.[34] It's time to roll up our sleeves and delve into how managing friction can help reduce energy waste.

Turbine engines have revolutionized power generation. Some turbines are aeroderivatives, meaning they're derived from the aircraft engines described earlier. The overall concept remains the same: compress air, mix it with fuel, burn it, and produce a hot gas. But the hot gas, instead of blowing out the back to provide forward thrust, rotates the turbine shaft, which is connected to a generator that uses electromagnetism to produce power. In the 1830s, Michael Faraday discovered that fluctuations in the magnetic field around an electrical wire induce an electric current. This process is called electromagnetic induction, since electricity is induced by magnetism. That same concept is used in power turbines. Magnets within the generator spin at high speeds, creating a magnetic field around an electrical wire. As the magnetic field attracts or repels the electrons in the wire, it creates a charge differential, pushing the electrons along the wire to generate a current.

Today, gas turbines can be coupled with steam turbines to create Combined Cycle Gas Turbines (CCGT), boosting their efficiency to over 60 percent. This is achieved by using the exhaust heat of the gas turbine as steam to power the aptly named steam turbine. The rotation of the steam turbine produces even more electricity than the gas turbine alone. The evolution from the gas turbine to CCGT power plants spanned nearly seventy years, spurred on by commercial rivalry be-

tween GE and Westinghouse Electric Corporation. The two companies
spent decades competing for a monopoly in the power industry. In
the 1950s, Ivan Rice, an engineer at GE, had an idea: he would capture
the waste heat of a gas turbine to help power steam turbines at a
customer's plant site. The plant had been running at 22 percent efficiency, and the customer needed to boost power output. Rice proposed
using a double-drum boiler, which consists of two long barrels stacked
on top of each other. The lower drum is kept at a lower temperature
and pressure to hold water, and the top drum holds steam. Some of
the steam in the top drum is sent to an additional steam turbine, where
it is further heated and pressurized. Meanwhile, the cooler steam,
which contains liquid particles, is funneled back to the water drum to
recycle the water. This cycling of heated water is more efficient than
trying to heat water in a single drum boiler. The successful design
improved the power output and reliability of the plant. Then, in the
late 1960s, a Westinghouse engineer figured out how to optimize
the pressure and temperature of the steam supply by combining
gas and steam turbines. This could be achieved by using the exhaust
heat of the gas turbine as steam to power the steam turbine. Decades
of optimization would follow, including better designs to connect the
turbines as well as the development of materials with improved
performance.[35]

As far as friction is concerned, the presence of steam adds a new
set of challenges. With all the extra moisture and condensation, lubricants, either solid or fluid, are necessary to prevent corrosion.
Lubricants not only must withstand the high temperatures of the
systems but also be designed so that their viscosity doesn't change as
temperature rises. Steam turbine oils have a viscosity index that indicates how much the viscosity is affected by temperature; the higher
the value of the index, the less susceptible to temperature the viscosity is. To determine the index, the viscosity of the oil is measured
at 40 °C and 100 °C, temperatures selected because they represent the
lower and higher end of typical engine operating temperatures. The
measured viscosity is then compared to two reference oils to create a
numerical rating scale.[36] Steam turbine lubricants typically have a

viscosity index of 100 or above, which is considered in the high range.[37] Indexes over 110 are considered very high. One way to achieve a higher index is by adding polymers to the base oil. These molecules break down at higher temperatures, improving the lubricant's temperature stability, which is key in maintaining the lubricant's internal friction across operating temperatures. Without high-temperature viscosity modifiers, the fluid film would break down, leading to dry contacts, high friction, and inefficient operation.

CCGT used in power plants offers over 60 percent efficiency and produces up to 50 percent more power than traditional power plants. Traditional coal-powered plants, on the other hand, operate at only about 33 percent efficiency. A major source of natural gas–supplied power generation, CCGT produces most of the electricity in the US, followed by coal-power plants and then nuclear plants.[38] As aging coal-powered plants are retired, they're being replaced with CCGT plants, putting us steps closer to solving the energy crisis. There is a catch, however. CCGTs are not a cure-all for energy savings. They produce about 350 kilograms of CO_2 emissions per megawatt hour from the combustion of natural gas. While this is half of what coal-powered plants produce, it needs to be reduced further to slow climate change.[39] One way to reduce emissions is through oxy-combustion systems that remove elements like nitrogen from the air, leaving behind highly concentrated oxygen. Oxygen is then used to burn natural gas, producing a by-product of CO_2 and water vapor. Water vapor can be condensed and separated from the CO_2, leaving pure CO_2. Compressing the CO_2 allows it to be stored and transported for other uses, such as making concrete and asphalt. If CCGT plants can operate with this system, then nearly all emissions can be captured. Research teams across the globe are working to scale up oxy-combustion attempt to achieve that goal.[40]

When it comes to the fuel used in power generation, natural gas has experienced a boom. According to the International Energy Association (IEA), the use of natural gas for power generation in the US increased by 7 percent in 2023.[41] Nearly half of all power generated in the US that year stemmed from natural gas. Compared to coal-power generation, natural gas produces less greenhouse gas emissions, which has been a key driver in its adoption. However, drilling for gas is an energy-intensive process. Some of the same issues with friction in manufacturing occur with the drilling machinery used in natural gas extraction. In this case, rather than a metal-cutting tool against a metal part, the drilling machine is cutting against earth. The amount of energy consumed in this process depends on the friction generated. Any materials used to reduce friction must be safe for the environment. Coatings, for instance, need to be formulated to interact with soils and rock while not leaving behind adverse wear debris. The lifetime of materials used in the coating must be at the forefront of design considerations. In Chapter 3, we learned that fluoropolymers can persist in the environment. Even though PTFE might perform admirably during drilling, it's unlikely to be the best choice for coating the outside of the pipes. Researchers have been exploring plant-based coatings that are safer for the environment, as well as materials derived from recycled products.

Friction between the pipe and its surroundings can also cause pipes to buckle. Take, for example, pipelines on the ocean floor, which at any time carry roughly 10 percent of the global supply of natural gas.[42] Some of these pipelines are long and narrow, which make them prone to movement. While some deformation can be tolerated, extreme deformation can lead to failure, particularly at connection points between pipes. A pipe's stability is tied to its axial stress, a force that acts perpendicular to the area of a body. It presses against or into the body, causing the material to stretch or compress. As pipes expand because of heat from the gas flow or the environment, they press against the surrounding soil. The friction between the pipe and the soil resists this expansion, constraining the pipe and building up axial stress. This can result in instability and buckling to release stress.

It might seem that the solution would be to minimize that friction. However, the same friction that increases axial stress can also help prevent the pipes from shifting about, or walking, when buckling occurs. Lowering the friction through material selection, surface finishing, or coatings can reduce the axial stress but may cause the pipes to move more. In 2022, researchers at the University of Bristol found a clever solution to the problem of buckling pipes. They designed pipe systems that intentionally buckled and moved at certain locations, minimizing the axial stress and expansion at the connecting points between pipes. To do so, they used surface texturing to create preferential high-friction zones. By placing these enhanced friction zones at various locations along the pipe, they found that positioning them at the pipe ends reduced expansion there by over 40 percent. Additionally, the location of the texturing gave them control over where buckling occurred, a clever way to control the flow along the pipes.[43] Intentional buckling may seem like an unexpected use of friction to help solve the energy crisis, but it's a reminder of the variety of clever ways we can use friction to achieve our goals.

Even with these friction-controlling approaches, pipelines and drilling remain energy-intensive processes. To truly tackle climate change and the energy crisis, we need our energy to come from renewable sources. The general approach may remain the same: something powers a turbine, which then generates electricity. That something could be natural gas flowing through the aforementioned pipelines, or it could come from renewable sources. Wind energy is the fastest-growing energy source in the US, generated by none other than wind turbines. These easily recognizable structures have caused all sorts of tribological headaches, not least owing to the sheer size of them. And wind turbines do not operate in ideal conditions for efficient machinery. Wind isn't constant, so trying to predict its fluctuation rate and magnitude is a challenge. At the end of the day, the most important consideration for wind power is reliability. If a wind turbine goes down, the productivity of the site decreases, increasing the cost of the resource and potentially causing nightmares for customers if there isn't a backup power source available. Gas and steam turbines are at least relatively easy to repair and maintain. They are easy to access, compared to wind turbines, which are often located offshore or in less accessible rural locations. When a wind turbine fails, the downtime to address its failure can be significant. Understanding and controlling friction can reduce maintenance intervals and improve both reliability and efficiency.

Gearboxes in particular can be a nuisance in wind turbines. Like the geared turbofan of an aircraft engine, a gearbox increases the rotational speed of the turbine. The wind turns the turbine blades at a relatively low speed. The blades are connected to a shaft that spins at the same rate, usually between thirty and sixty revolutions per minute. This slow-spinning shaft connects to the gearbox, which then spins an output shaft at around a thousand revolutions per minute. Inside the generator are coils typically made of copper and magnets. In some turbines, the magnets spin at the same rate as the shaft. In others, the copper wires do. The spinning generates a magnetic field around the copper. This magnetic field exerts a force on the copper electrons, driving them through the wire and generating electricity. The faster the coils spin through the magnetic field, the more electricity is produced, which is why the gearbox is present—to speed up the shaft rotation. This enables the generator to convert mechanical energy to electricity.

Thanks to friction, the gearbox is responsible for the bulk of the mechanical energy losses in wind turbines. For a five-megawatt wind turbine, viscous drag losses in the gearbox can reduce output by as much as 3 percent.[44] (One megawatt is equivalent to one million watts.) These energy losses limit generator efficiency. Synthetic oil lubricants can withstand the temperature and loading conditions of the gearbox, but over time, they will need to be topped up. The inaccessibility of wind turbines makes this a time- and cost-intensive process. But without maintenance, the lubricant will fail. The ramifications of lubricant failure and maintenance challenges are leading researchers to think outside the (gear)box.

Some tribologists have proposed radically new designs, such as permanent magnet synchronous generators (PMSG), which bypass gearboxes altogether. In traditional wind turbine designs, spinning is required to produce a magnetic field and generate current. Permanent magnets, such as the rare-earth metal neodymium, can generate the equivalent magnetic field to produce a current at the lower shaft speed, eliminating the gearbox and the frictional losses associated with it. To become viable, PMSGs would need material innovation, as they require expensive rare-earth metals for magnetization. When this problem is solved, however, they will not only offer the advantage of reliability and efficiency over the gearbox design but also help with an issue that wind turbines share with jet turbines: maintaining efficiency at different speeds. At low wind speeds, wind turbines struggle to generate enough rotational speed to produce power, experiencing as much as a 90 percent reduction in efficiency. According to the Quad team, if this challenge can be met, another quad of energy could be freed up from wind turbines alone.[45]

In our efforts to reduce energy, friction is indeed a powerful tool. After all, if there's one thing we've learned from tribology, it's that friction is always present. We've seen numerous ways in which recognizing its presence and understanding its role in an application can significantly reduce energy usage and emissions. Sometimes, such as with large, noisy machine parts, it's obvious that we need to reduce friction as much as possible. But other times, such as with soil and pipelines, friction demands more finesse, requiring a keen eye to detect it and the expertise and patience to balance it. This is why the Quad team recommended that tribologists always be included in energy audits. In some countries, companies can acquire energy ratings for their facilities. Leadership in Energy and Environmental Design (LEED) certification is perhaps the most well known internationally, setting standards for energy-efficient and sustainable buildings. In the United States, companies can pursue a Superior Energy Performance certification through the Department of Energy's Industrial Assessment centers. These centers review a company's energy consumption practices and make recommendations for adopting more sustainable practices. An independent team then verifies the company has met certain energy standards. The audit teams are usually multidisciplinary but almost always lack an expert in friction. They don't even mention friction or tribology in any of their energy assessment training materials. By educating those seeking certifications and those conducting audits about the impacts of friction, we can effect large-scale energy savings across a variety of sites, from offshore wind farms to manufacturing floors.

While we have come a long way in understanding tribology and pushing the boundaries of research, we still have work to do in raising awareness of friction. That might be as simple as including tribology in training manuals for energy audits or in government energy-saving targets. In recent years, engineering departments at universities have begun to include tribology courses in undergraduate catalogs, but this is not the norm across the country, let alone around the world. In the decade since completing my PhD, I've witnessed more tribologists being hired at universities and running labs. This has already resulted in more specialists and awareness of friction and wear. There is power in having more people recognize friction and understand how it works. It is thinking with a friction-forward mindset that will unlock the twenty quads of energy the Quad Team estimated tribology can save. You never know when one of those "Oh, that's friction!" moments may lead to the next innovation, like Teflon, or an unexpected part design, like a wind turbine without a gearbox.

Or, perhaps, friction will help us unlock other scientific mysteries.

6 *Frontiers of Friction*

AS I WAS WRITING THIS BOOK, my mind was inundated with examples of tribology. Florida is a playground not just for tribology but for extreme engineering. Yet one example stood out: the space shuttle launch from Cape Canaveral, home of the Kennedy Space Center, which carried miniature tribometers made in our lab. It may come as a surprise that there is friction in space, given that most of space is a vacuum. But space is a dynamic place, where planets orbit around one another, occasionally veering off course. Galaxies collide and black holes swallow up stars. Celestial objects may not experience visible rubbing, but they do move relative to each other. And where there is motion, there is dissipation of energy, or friction.

As our understanding of friction has expanded, so too has our appreciation for friction's influence on the physical world at its smallest and largest scales. From the quantum forces acting on fundamental particles of matter to the gravitational forces holding together galaxies, friction is everywhere. This perhaps explains why tribology is at the heart of so many fields, including astrophysics, quantum mechanics, and molecular biology, the topics of this final chapter.

Tidal Friction and the Evolution of Tidal Theory

In the 1690s, the English astronomer Edmund Halley noted that the Moon appeared to be speeding up as it moved across the night sky. At the time, Halley concluded, incorrectly, that the Moon must be moving closer to Earth. The opposite turns out to be true; the Moon is moving away, and as a result, Earth is slowing down. To understand Halley's error, we must look to the waters of our own planet.

For millennia, philosophers and astronomers sought to explain the ocean's tides. Aristotle believed the tides were caused by wind produced by the Sun's heat. As early as 150 BC, astronomers associated tides with the Moon. The Greek astronomer Seleucus connected the height of the tide to the Moon's position in the sky. With the fall of the Roman Empire, the disruption to learning and knowledge halted progress on tidal theory in the West. Fortunately, Eastern scholars continued to pursue the topic. In the thirteenth century, Persian astronomer Zakariya al-Qazwini taught that moonlight warming the water caused the swells of tides. However, it wasn't until the seventeenth century, when Johannes Kepler suggested that the tides were the result of the Sun and the Moon pulling water toward them, that modern tidal theory began to take shape. Kepler was unable to prove his theory, and it would be Isaac Newton who finally solved the puzzle.[1]

In *Principia*, Newton proposed that as the Sun and Moon pull on Earth through gravitational attraction, a tidal force arises, causing the ebb and flow of oceans. To understand his explanation, we first need to look at how the Moon's gravity affects Earth. Points on Earth nearest to the Moon are pulled toward it with a stronger force than points farther from the Moon. This differential gravitational effect is called the *tidal force*. It stretches and compresses Earth along the Earth-Moon axis, causing water to bulge in the direction of the Moon. These bulges are sometimes called tidal bulges. The part of Earth rotating through the bulge will experience high tide. However, we know that there are two high tides, roughly twelve hours apart, which means there's another bulge on the opposite side of Earth. The second high tide is the result of inertia generated by the rotation and motion of the Earth-Moon system. As Earth rotates, water moves away from the Moon due to inertia. On the far side of Earth, where the Moon's gravitational force is weaker, inertia overcomes the effects of gravity, causing water to bulge in the direction opposite of the Moon. As each part of Earth's surface rotates through this second bulge, it experiences a second high tide. Between these high tides are low tides, generated as Earth rotates out of the bulge.[2] Although we see the effects in our oceans, we don't feel tidal forces because our masses are infinitesimally small compared to those of Earth and the Moon.

The Sun also exerts a gravitational force on Earth. This force is weaker despite the enormous mass of the Sun due to the greater

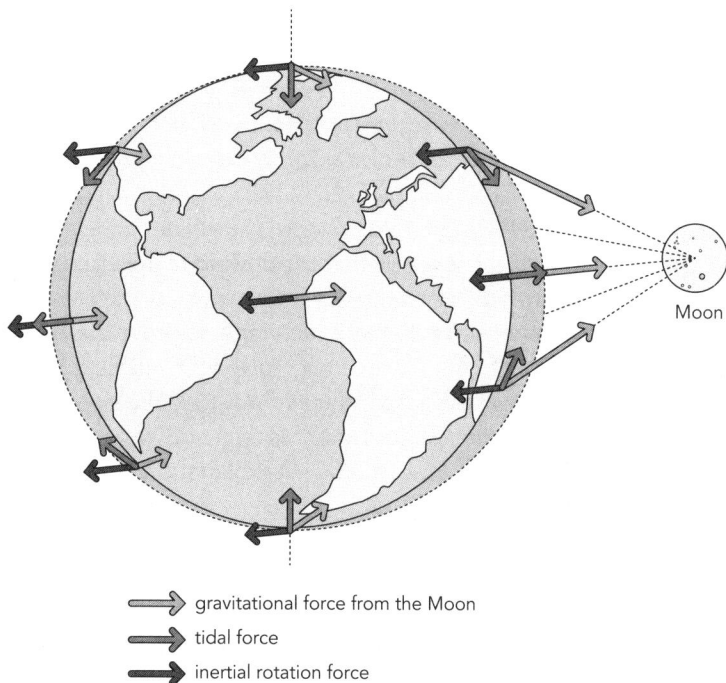

gravitational force from the Moon
tidal force
inertial rotation force

Tidal forces acting on Earth

distance between the Sun and Earth. However, we notice the effect of solar tides when Earth, the Moon, and the Sun align. The tidal forces generated by the Moon and the Sun become superimposed, giving us a higher-than-normal high tide. Known as a spring tide, the phenomenon was first documented around 325 BC by the Greek explorer Pytheas as he sailed around the British Isles. The tides that we experience are the cumulative effect of the forces generated by both the Sun and the Moon.

In the nineteenth century George Darwin, the son of Charles Darwin, connected the tides to Halley's observation. After studying math at Cambridge, Darwin was awarded a prestigious fellowship to study at Trinity College. Like many Cambridge graduates at the time, he chose to study law and was admitted to the bar. However, he never practiced law. Instead, he pivoted back into the sciences, returning to Cambridge to begin his fellowship. At first, he investigated the biological impacts of first-cousin marriages, in part to determine if his own parents' marriage had led to the poor health and deaths of three of their children.[3] But ultimately, he was drawn to geology and astronomy.

Darwin wanted to understand the origins of the Earth-Moon system. He believed that the ancient, molten Earth had at one time been spinning so fast that a part of it broke off and was thrust into space, becoming our Moon. He called this "fission theory." To prove it, he showed that the rapid motion of Earth would create a force massive enough to overcome the planet's gravitational pull. A glaringly obvious piece of evidence against this theory, however, is the fact that our planet has only one moon. If fission theory were correct, the odds are slim that only one satellite would have formed. Nonetheless, fission theory became the accepted theory of Earth-Moon evolution. It would take the Apollo Moon missions to disprove it. When dated, lunar samples collected by Apollo astronauts were significantly older than what had been predicted by fission theory, and they appeared to be sourced from two different molten materials, as opposed to just one from Earth.[4]

Though fission theory was wrong, it provided key insights into the influence of friction on the Earth-Moon system. To prove that molten material could detach from Earth, Darwin had to consider all forces acting on Earth's molten material, including tidal forces. He realized that friction must be slowing down the tides. If tides were frictionless, high tide would occur when the Moon reached its zenith, or when it was directly overhead, in its closest position to Earth. That's not the case. High tide occurs slightly before this mark. Darwin believed that tidal friction, the friction caused by matter moving as it's pulled on by tidal forces, could slow the ocean enough to explain this phenomenon. On Earth, tidal friction includes friction from the seabed floor, drag from land interactions, and the internal friction of water itself. And this, Darwin realized, was key to explaining Halley's observation.

As water is pulled toward the Moon, the motion of the water is offset by the tidal friction slowing it down. This offset creates a twisting force known as a torque that acts against the rotation of Earth. In other words, the offset tides act as drag, slowing down Earth's rotation. Darwin showed that this rotational deceleration affects the Moon because of a physics principle discussed in Chapter 2: conservation of momentum. In the case of rotating bodies such as the Earth-Moon system, the type of momentum conserved is known as angular momentum. *Angular momentum* is a measure of an object's rotational motion, determined by the product of its mass, velocity, and distance from the center of its rotation. As Earth loses angular momentum from tidal friction, the Moon must experience an increase in angular momentum. This pushes the Moon into a higher orbit, farther away from Earth. It also explains why Halley observed the Moon to be accelerating.[5]

Darwin's quantitative proof of the impact of tidal friction on the Earth-Moon system is considered classic work in tidal dynamics. In 1968 American oceanographer Walter Munk quantified the components of tidal friction. He estimated that 60 percent of tidal energy is dissipated by friction on the seabed floor, 20 percent by waves, and 3 percent from the stretching of Earth by the Moon's gravity. While the exact numbers are still debated, by understanding the mechanisms underlying components of Earth's tidal friction, we can predict when the Earth-Moon system will reach steady state, that is, when the Moon will stop moving away from Earth and Earth's rotation will cease to decelerate.[6]

When the Earth-Moon system reaches a steady state, the two bodies will become tidally locked. *Tidal locking* is a phenomenon in which a body's orbital period is the same as its rotational period. Earth isn't yet tidally locked to the Moon, but the Moon is already tidally locked to Earth. It takes about the same amount of time—a month—for the Moon to spin once around its axis as it does for the Moon to complete its orbit around Earth. This explains why we only ever see one face of the Moon. The so-called far side of the Moon was last visible from Earth billions of years ago, before tidal locking occurred.

What causes tidal locking, and how is it related to tidal friction? The Moon experiences its own tidal bulges from Earth's gravity. They're less noticeable since water isn't rising or falling, but they're there. Early in the evolution of the Moon-Earth system, the Moon experienced offset bulges that, due to the rotational differences between the two bodies, caused the same deceleration that Earth experiences today. The Moon decelerated until its tidal bulges aligned with those of Earth and it became tidally locked to our planet. All bodies that experience tidal forces decelerate until they experience minimal torque and their rotations become stable. In about 50 billion years, Earth will become tidally locked to the Moon, at which point one side of Earth will never see the Moon. That is, of course, if Earth can somehow avoid being destroyed by the death of the Sun five to ten billion years from now.[7]

Tidal locking is found throughout the universe. In our solar system, Mercury is tidally locked to the Sun, and the large moons of the planets are all tidally locked to their host planets. Theoretical work has suggested that planets in the habitable zone of their host stars may already be tidally locked. Often, the main indicator of habitability is the distance of a planet from its host star. To be habitable, a planet would need to be close enough to its central star to attain the temperatures necessary to sustain liquid water: 0 °C to 100 °C. The proximity to the central star would, in turn, increase the gravitational pull on the planet, causing tidal locking.

A tidally locked planet in the habitable zone would likely experience stark temperature differences between the side facing the central star—the permanently light side—and the side facing away from the star—the permanently dark side. This raises the question of how life could survive on such a planet. Life needs light for energy. But the light side would reach such scorching temperatures that DNA and RNA would break apart. Some researchers have proposed that there could be a sweet spot between the light and dark sides, where conditions are more favorable to life. On the dark side, the frigid temperatures might generate fierce storms as the atmosphere sucks in hot air from the light side, creating a literal twilight zone.[8] NASA is planning an

$11 billion Habitable Worlds Observatory, expected to launch in 2040, to search for potentially habitable exoplanets. Among the features of interest for this telescope will be the presence of tidal locking and twilight zones. Of course, even if such planets are identified, they'll need the magic combination of water, temperature, and atmosphere that scientists hope the next generation of telescopes will help identify.

Where There's Friction, There's Heat

Temperature, as we have seen throughout this book, is often in conversation with friction, including tidal friction. The slowing rotation of Earth or another satellite is caused by the dissipation of energy that results from friction. This mechanical energy is converted into heat, a process called *tidal heating*. Today, tidal heating on Earth isn't significant, producing only about two-and-a-half to three-and-a-half terawatts of heat.[9] To put that in perspective, it would take over a thousand days for the thermal energy of tidal heating to equal the power consumption of the United States alone. Earth has had a long history, however, and tidal heating was not always insignificant. When the Earth-Moon system was forming 4.5 billion years ago, tidal heating may have shaped the planet as we know it.[10]

Today, the leading theory of the Moon's formation is the giant-impact theory. According to this theory, first proposed by William Hartman and Donald Davis in 1974, a large collision between early Earth and another protoplanet generated large amounts of debris that coalesced into the Moon. This likely happened close to Earth, about seventeen times closer than the Moon is today, producing short five-hour days. With its proximity to Earth, the Moon had less momentum and rotated more slowly compared to today. Earth, on the other hand, had more momentum and rotated faster. Tidal energy pushed the moon into a stable orbit and eventual tidal locking. The resulting tidal heating produced enough thermal energy to maintain the flow of Earth's magma for a couple million years. Once the Moon migrated sufficiently far from Earth, the tidal heating effect was no longer strong enough to maintain the planet's molten material. As Earth cooled, the mantle, the primarily silicate layer between Earth's heated core and crust, began to form. Gases and vapors were released, producing water and steam and establishing a carbon dioxide–rich atmosphere. Organisms such as algae evolved to feed on the carbon dioxide, releasing oxygen into the atmosphere and beginning the process of creating an oxygen-rich environment capable of supporting complex life.

Would life have emerged earlier without tidal heating? We'll likely never know since tidal heating was synchronous with Earth's development, and the thermal evolution and composition of the mantle was inextricably tied to friction's effects.[11] A lack of concrete data makes it challenging to reconstruct the evolution of Earth during the 700 million years from its formation to the cooling of its mantle. Another challenge with answering this question was first described by astronomers Carl Sagan and George Mullen.

Evidence from zircon rocks suggests that there may have been liquid water on Earth over four billion years ago. The problem is that at the time, the Sun was only 70 percent of its current intensity and wouldn't have been able to warm Earth enough for liquid water to exist. This is called the Faint Young Sun paradox. Some explanations for this paradox have focused solely on the atmosphere. But recently scientists have shown that tidal heating might have raised the temperature of Earth by several degrees. This, combined with the temperature effects from greenhouse gases, could have prevented Earth from freezing over before the Sun was strong enough to do so.[12]

Whereas physical evidence of the evolution of Earth and the Moon is scarce, the moons of the outer planets in our solar system are rich in clues about the impacts of tidal friction on planetary formation. To date, astronomers have identified ninety-five moons orbiting Jupiter. Three of these moons—Ganymede, Europa, and Io—have orbital patterns that result in impressive tidal friction. Their orbital periods are related by simple ratios, called orbital resonance. For every orbit of Ganymede around Jupiter, Europa will orbit twice, and Io four times.[13] Jupiter, Ganymede, Europa, and to a lesser extent the moon Callisto, all pull on Io's solid material. In 1970 researchers suggested that the tides induced by Jupiter were large enough to generate tidal heating on Io, a solid moon the size of our own. Shortly after the study was published, the spacecraft *Voyager I* sent back photos of volcanic activity and molten rock on the distant moon, evidence of frictional melting. When melting on Io occurs, heat needs to escape somehow, resulting in one of the most geologically active bodies in the solar system. The *Galileo* spacecraft that orbited Jupiter from 1995 until 2003 took dazzling photos of lakes and rivers of lava. In one of these images, the mission even captured a dramatic eruption of volcanic plumes rising above the horizon. Io's landscape, scientists believe, is likely a mirror of what Earth looked like billions of years ago, as tidal friction maintained the planet's magma oceans.[14]

Sometimes, the molten material induced by tidal heating isn't magma but rather water. Many science fiction movies, TV shows, and books have focused on Europa, with its seemingly mystical subsurface ocean. This ocean, and its proximity to the icy moon's silicate core, is a perfect recipe for life and habitability. Europa, and Saturn's moon Enceladus, are considered the most probable sources of life beyond Earth in our solar system due to evidence of oceans beneath their icy exteriors. Tidal friction provides the heating necessary to create liquid water under a thin shell of surface ice. The upcoming launch of the *Europa Clipper* is expected to further advance our understanding of tidal friction and tidal heating, providing the closest look at Europa to date. The mission will help to determine how much heating is from the radioactive decay of Europa's mantle and how much is from tidal heating. That data can then be used to determine the magnitude of tidal friction, elucidating how tidal friction has shaped Earth. Information from the mission may ultimately inform the search for habitable exoplanets.

In the search for habitable exoplanets and how tidal heating may influence them, some researchers have become interested in the TRAPPIST-1 system. Its dwarf star, less than 10 percent of our Sun's mass, was discovered in 1999, and its planets were first observed in 2016. In total, this compact system has seven planets that are in orbital resonance and likely tidally locked. The central star may be cooler than ours, but the planets' close orbits leave a handful in the habitable zone. As I write this, researchers are applying tidal friction theory to understand the planets' orbital mechanics as well as their potential for tidal heating. Tidal heating may be generating enough interior heating to yield volcanic activity on Io and a hidden ocean under a layer of ice on Europa.[15] When future missions investigate the moons of our own outer planets, their insights into the impact of friction on the orbital mechanics and thermal evolution of these celestial bodies will significantly expand our understanding of potentially habitable environments beyond our solar system.

Dynamical Friction

Space is an unfathomably large vacuum, dotted with celestial objects such as planets and stars, all in constant motion as the universe expands. With the exception of asteroid collisions, celestial bodies don't tend to come in contact with each other. Despite this, we've already explored one example of friction at play in the cosmos. Through tidal forces, gravitational pull can create an important frictional phenomenon. Tidal friction concerns the matter *within* bodies, such as oceans or molten material. Gravitational forces can also act like drag on an entire body as it moves through space, resulting in energy dissipation. This is known as *dynamical friction,* not to be confused with *dynamic* friction, first mentioned in Chapter 1.[16]

Dynamical friction is the loss of momentum and kinetic energy owing to gravitational interactions as objects move through space. It is, quite appropriately, also referred to as gravitational drag. In the 1920s and 1930s, scientists realized that the universe is composed of galaxies of various shapes that continuously rotate, collide with, and swallow up other galaxies. These galaxies contract, and also push away from each other at increasing speeds. They wanted to understand this motion and determine if or how equilibrium, a state in which the system is no longer expanding or contracting, occurred. Subrahmanyan Chandrasekhar was one such scientist who devoted his energies to the topic. Today, he is widely credited with launching the field of modern astrophysics.[17] In a wonderful etymological touch, he would go by "Chandra," which means "luminous" or "moon" in Sanskrit.

While many of the scientists introduced in this book came to study friction from a background in engineering, driven by a practical need to manipulate friction, Chandra had a different motivation. One of the twentieth century's eminent astrophysicists, he was among the first to combine physics with astronomy. He doesn't have a variable named after him like Coulomb, but he does have an equation, a telescope, an X-ray array, and an asteroid that all bear his name. Chandra was also awarded the 1983 Nobel Prize in Physics, making him only the third person to win it for astronomy or astrophysics. In his book *The Demon-Haunted World,* Carl Sagan thanked Chandra for helping him discover "mathematical elegance."[18]

Born in 1910 in Lahore, which at the time was part of British India and is now in Pakistan, Chandra would later settle with his family in Madras and then spend much of his career at the University of Chicago. He received his bachelor's degree in physics from Presidency College in India at the age of nineteen. Upon graduating, Chandra was awarded a government scholarship to pursue graduate studies at Cambridge University. During his voyage from India to England, the nineteen-year-old Chandra did much of his groundbreaking work on white dwarf stars, determining that they reached a critical mass before collapsing into a neutron star or black hole. At the age of twenty-one, he published this work, and eventually, this mass limit would become known as the Chandrasekhar limit.

Chandra continued working on the structure and evolution of stellar systems through the 1940s. He then put it away, returning to it later in his career. As Chandra explained, he would find a topic or area that interested him and work devotedly on it until he felt able to lecture and write a treatise on it, at which point he'd move on to another topic. Chandra organized science into two general categories: basic and derived. Basic science was concerned with analyzing matter and concepts of space and time. Derived science sought to understand and explain natural phenomena in terms of basic science. He would give astronomer Edmund Halley's identification of his eponymous comet as an example of basic science and the application of Newton's gravitational law to describe the comet's orbit as an example of derived science. Derived science was Chandra's passion. It married his strengths in physics and mathematics beautifully. This passion led to the discovery that objects moving through space experience drag, which he named dynamical friction.[19]

After exploring the evolution of stars, Chandra sought to understand how stars move in clusters and galaxies. He took a statistical approach to the problem. Statistical models describe the likelihood of an event occurring, such as rolling a three on a die. Chandra used statistical modeling to describe how clusters of stars might interact with each other. He found that if the gravity of passing stars interacted with each other, the stars would slow down and even collide.

In 1942 he published the results of his efforts, the now-classic book *Principles of Stellar Dynamics*. In it, Chandra estimated how long it would take for these interactions to cause collisions. This work was done with the characteristic mathematical rigor and elegance that Sagan was later in awe of; however, the resulting velocities of the objects in the stellar system left open some questions. According to the estimate Chandra had produced, in the absence of some outside event, the average velocity of a cluster of stars would increase indefinitely. Today, we know that this can and does happen because the universe is expanding. However, the behavior of galaxies and star clusters spiraling inward toward their center due to gravity contradicts boundless velocity growth. Chandra knew that another factor must be at play, causing such behavior.

In an article published later that year, titled "New Methods of Stellar Dynamics," Chandra explained that the missing piece was the dissipation of energy and momentum due to dynamical friction.[20] A massive object exerts a gravitational force that pulls the smaller objects toward it, creating a wake of smaller objects, much like the waves of water created by a boat. This dense wake acts as drag, slowing the massive object down. It explains why larger stars tend to be closer to the center of a star cluster. Such drag is not exactly the mechanical rubbing we traditionally associate with friction. However, the net effect is a loss of kinetic energy. Instead of interatomic interactions, the source of friction is gravitational interactions. Since, as Chandra found, the energy loss of the large object is proportional to its velocity squared, dynamical friction has less of an effect on objects moving at very high speeds. Intuitively, this makes sense. If an object is speeding through a galaxy, there's less time for it to be affected by the wake.

This groundbreaking work might never have come to fruition, had Chandra allowed an early clash with a titan in the field to derail his career. Arthur Eddington was well known for his work on the luminosity of stars and for introducing Einstein's general theory of relativity to the English-speaking world. He had assumed the elite position of Plumian Professor of Astronomy and Experimental Philosophy at Cambridge University after the unexpected death in 1912 of George Darwin, who had last held the post. Eddington had been one of Chandra's assessors at his doctoral defense, and also a colleague at Trinity College, Cambridge, where Chandra received a fellowship upon completing his doctoral studies. Eddington suggested Chandra present his paper on the evolution of white dwarves at a Royal Astronomy Society meeting in 1935. To Chandra's surprise, Eddington then publicly derided his work at that meeting, telling the room that Chandra's paper was just mathematics and impossible since stars couldn't simply collapse in on themselves. In two other public speaking events, Eddington would again disparage Chandra's work. It turned out that Chandra's theory was in direct opposition to a theory that Eddington had worked on, showing that white dwarves did not collapse.

Having someone of Eddington's stature ridicule an early scientist's work can be career ending. It was a huge blow to Chandra. Privately, Chandra asked other renowned physicists to assess his work. They concluded that Chandra was correct, although most were unwilling to say so publicly. One physicist, Gerard Kuiper, after whom the Kuiper Belt in our galaxy is named, did speak out, proclaiming that his own work supported Chandra's theory. Eventually, other physicists would publish work confirming Chandra's results and calling Eddington's attacks a "misunderstanding." That Eddington, given his stature, did not end Chandra's career is perhaps miraculous and a testament to

Chandra's perseverance. Had Chandra yielded to Eddington's attacks, or even simply retreated from the forefront of science, a trove of critical astrophysics work might never have come to fruition. Yet despite these clashes, the two remained admirers of each other's work, and Eddington co-sponsored Chandra's 1944 nomination for the prestigious Fellowship of the Royal Society. In his eulogy to Eddington after his death, Chandra called Eddington one of the greatest astronomers in history.[21]

Friction and Dark Matter

The visible matter we can see, such as the star clusters Chandra studied, makes up only a small percentage of the mass of the universe. It's estimated that atoms and light constitute as little as 5 percent of the universe. The rest is dark matter and dark energy. Dark matter, called "dark" because it doesn't emit or absorb light and cannot be seen, is estimated to make up about 27 percent of the universe. It was theorized in the 1930s by the Swiss astronomer Fritz Zwicky. Now regarded as "the father of dark matter," Zwicky was studying the Coma Cluster of galaxies, located 300 million light years away, when he homed in on some curious behavior. At the speed the galaxies were traveling, they shouldn't have been able to stay intact. Perhaps, he proposed, some unobserved mass was holding them together. Zwicky called this undetectable material *dunkle Materie,* or dark matter.[22]

The gravitational pull of dark matter may explain why physics-defying galaxies don't fall apart. Dark matter may also explain why essential elements for life, like carbon, oxygen, and nitrogen, are able to accumulate in galaxies, leading to the formation of habitable, rocky planets, rather than being ejected and floating off into space. To prove such ideas, physicists must first detect and measure dark matter. The problem is that dark matter doesn't absorb, emit, or reflect electromagnetic radiation such as light, radio waves, or X-rays.

In Geneva, Switzerland, scientists at CERN's Large Hadron Collider have been smashing together particles at unimaginably high speeds in order to detect a loss of energy and momentum, which might indicate the presence of a dark matter particle. Since energy and momentum are conserved and can't be destroyed, any loss would indicate that *something* invisible is present to which energy and momentum have been transferred. These experiments must be run at cryogenic temperatures barely above absolute zero to eliminate thermal noise from measurements. Scientists have yet to directly detect dark matter, but they have detected incredibly small masses, the equivalent of one-fifth of a proton. For now, dark matter remains elusive. However, this may soon change. As more scientists have joined the search for dark matter, a promising approach, based on gravitational interactions and dynamical friction, has emerged.

If Zwicky was correct about dark matter binding galaxies together, then dynamical friction affects more than the galactic matter we can see. Galaxies come in various shapes and sizes. Larger galaxies often have smaller satellite galaxies orbiting around them. These galaxies interact in a variety of ways. At the extremes, they can collide and merge together. But there may be another way—through dynamical friction. Dark matter forms a halo around galaxies. These so-called dark matter halos have been inferred by the otherwise inexplicable behavior of light and the gravitational interactions of visible matter. Sometimes, a smaller satellite galaxy can enter the influence of a larger host galaxy. As the smaller galaxy moves through a dark halo, dark matter absorbs its energy and momentum, causing it to slow down and orbit closer and closer to its host galaxy. This is known as orbital decay, or a decaying orbit. Dynamical friction is greater in high-density regions where there is more dark matter to absorb the satellite's energy and momentum. Ultimately, orbital decay can lead to a phenomenon known as satellite sinking, in which a satellite sinks into and merges with a larger galaxy.[23] To estimate the decay rate, scientists use Chandra's formulas for dynamical friction, which calculate how long the deceleration will take.[24] With new technology such as improved gravitational wave telescopes providing higher-quality data, they're now able to observe satellite galaxies in the Milky Way and develop rigorous modeling of their decaying orbits.

Satellite sinking is a form of accretion, the process by which matter comes together under the influence of gravity. Accretion is at the heart of the structure of the universe. It is responsible for the growth of black holes, dense cosmic objects with such strong gravitational pull that not even light can escape from them. One of the most intriguing behaviors of black holes is that they consume or swallow nearby objects, such as stars. Just as dynamical friction has helped determine galaxy decay rates in satellite accretion, it can also be used to determine the decay rates of stars accreted by black holes.

Using Chandra's formulas for dynamical friction, a team of researchers at The Education University of Hong Kong (EdUHK) discovered the first indirect evidence of dark matter surrounding black holes. The clue to the significance of dynamical friction came from the orbits of two stars: XTE J1118 + 480l, located in the Ursa Major constellation, and A0620–00, in the Monoceros constellation. These stars were companion stars to two black holes that the team had been studying. The stars' orbits were decaying, meaning that each star would eventually collide with its respective black hole, which would consume it, growing through accretion. The rate of decay was about one millisecond per year. One millisecond per year represents a completely insignificant amount of time for most of us, but to the team, this was a sure sign that friction from dark matter was present. According to the team's calculations, the stars' orbital decay should have been fifty times smaller. The researchers proposed that if dark matter was present, the decay would be faster than predicted since dynamical friction would drag the star closer to the black hole. So, off to the computers they went, simulating the star–black hole companion systems with the inclusion of dynamical friction from dark matter.[25]

Their theoretical results matched the observed decay rate. Their experiment has been a game changer in the search for dark matter. Dark matter researchers no longer need to rely on gamma rays, high-energy radiation of the shortest wavelength, to detect the energy emitted from collisions of black holes—not a particularly common event. Instead, they can combine observed orbital rates and dynamical friction. This might just be the tool we need to detect an elusive type of black hole that could unlock mysteries about the origins of the universe.[26]

Dynamical Friction and Primordial Black Holes

When I first learned about black holes, I felt both awestruck and terrified. We're taught that these ominous-sounding cosmic objects, from which even light cannot escape, form when stars collapse, causing a supernova explosion. But that may not actually be how all black holes form. In 1966 two Soviet astrophysicists, Yakov Zel'dovich and Igor Novikov, proposed a radical idea. They suggested that black holes formed within moments of the Big Bang, well before any star had time to collapse.[27] Known as primordial black holes, these celestial bodies, if they exist, would be older than the stars. At the time of Zel'dovich and Novikov's proposal, it was believed that these black holes would be enormous because black holes grow by consuming the matter around them. When scientists failed to find evidence of primordial black holes, interest in Zel'dovich and Novikov's proposal waned. But not for long.

Shortly after Zel'dovich and Novikov's publication, Stephen Hawking wondered if perhaps the opposite was true: that the older the black hole, the smaller it would be. He proposed that the newborn universe contained pockets of subatomic particles packed so densely that they had collapsed in on themselves from gravity. If this was the case, then the resulting black holes would be small. Very small.

Hawking modeled these black holes from a particle physics perspective. He considered pairs of particles with opposite charges, called particle-antiparticle pairs. When these two types of particles met, they'd annihilate each other. But, if they met at the edge of a black hole, interesting things could happen. One particle might fall into the black hole while the other managed to escape. The escapee particle would be emitted as faint electromagnetic radiation, called Hawking radiation, leading to the evaporation of the black hole over time. Given the age of primordial black holes, those that still existed could be on the scale of micrometers. Talk about trying to find a needle in a haystack!

Though still unproven, Hawking's work on primordial black holes made significant progress toward bridging general relativity, quantum mechanics, and classical thermodynamics—the holy grail of physics. Detecting these black holes not only would validate this work but might even lead to new concepts that unite these fields definitively. And by dating and studying the composition of these black holes, scientists might finally be able to answer some of the most bedeviling questions about the early universe, such as when and how stars formed, and how they evolved into galaxies and solar systems. If primordial black holes exist, stars could have started forming hundreds of millions of years earlier than our current theory predicts.[28] Of course, detecting micron-sized stellar objects in the vastness of space is no trivial task. However, when large, asteroid-sized primordial black holes interact gravitationally with surrounding objects, dynamical friction will occur. And as the EdUHK team demonstrated, this dynamical friction could be a powerful tool for detecting them.

In addition to answering questions about the universe's origins, confirming the existence of primordial black holes could eventually shed light on the nature of dark matter itself. Primordial black holes may consist entirely of dark matter. Some scientists have even speculated that primordial black holes *are* dark matter. According to this theory, primordial black holes formed early enough after the Big Bang that the elements in visible matter, and the chemical reactions required to create it, hadn't emerged yet. Of course, to validate or debunk this idea, we must be able to detect primordial black holes.

If primordial black holes exist, their presence may have significantly influenced the formation of the universe. As stars began to form around dark matter clusters, they may have been slowed down by dynamical friction and pulled inward. If so, then primordial black holes might have developed into normal black holes through accretion. Their evaporation, and the energy and particles that radiated into space as a result, would have played a role in the processes and reactions that formed regular matter. Even today, it's possible primordial black holes are still evaporating. They might be the source of cosmic rays that speed through the universe at nearly the speed of light, striking planets and other bodies in space, scattering subatomic particles, and catalyzing reactions like ionization from their radiation.[29] And despite being so small, primordial black holes could potentially influence nearby stellar structures by altering their trajectory through space.[30] Some primordial black holes are believed to be so dense that they can exert enough of a gravitational effect to influence the orbit of a planet the size of Mars.[31]

The influence of primordial black holes on the formation of stars is an intriguing question. Hawking believed it would be possible for a neighboring star to capture one of these black holes. This scenario could happen during the star's formation. The small size of the primordial black hole would keep it from destroying the developing star. Instead, the black hole would sink to the star's center. There, the black hole would grow slowly since the energy of the star's nuclear fusion would overpower and neutralize the black hole's accretion.

Such a scenario was recently simulated by scientists at the Max Planck Institute for Astrophysics. The team found that the resulting so-called Hawking star could survive hundreds of millions of years and would be nearly indistinguishable from other stars except for its core temperature. A Hawking star's core would be cooler since black holes absorb all energy, including heat.[32] But while the core temperature of a star may be an indicator of a black hole's presence, it can't tell the full story. Probes have detected hundreds of giant stars with cooler than expected cores, but this isn't necessarily the effect of a black hole. It could also be related to the age of a star, a sign that the core is cooling as it ages. Scientists must therefore turn to other detection methods to seek out Hawking stars. In a Hawking star, hotter layers would radiate outward from the cooler black hole as it evaporated. The frequency of the convective energy would differ from the energy of a normal star, whose dense plasma core is powered by nuclear fusion. For this reason, scientists have turned to astroseismology to hunt for these theoretical stellar objects.[33] Astroseismology involves using acoustic oscillations to measure the structure of stars. It may eventually uncover a black hole at the core of a star by comparing the star's acoustic signature to that of a typical star, like our Sun. The

European Space Agency's PLAnetary Transits and Oscillations of Stars (PLATO), expected to launch in 2026, aims to do exactly that as part of its mission.[34]

PLATO is just one such project. Other efforts to detect primordial black holes involve probes equipped with gamma ray detectors and telescopes capable of detecting lensing effects. Gravitational lensing occurs when celestial objects produce a curvature in the spacetime continuum, causing light to bend as it passes around them, as if distorted by a lens. Telescopes can pick up very small distortions in light and gravitational interactions between celestial bodies that can't be readily explained.[35] Gravitational lensing may eventually offer a way of detecting primordial black holes, although the technical difficulty of catching the lensing of such a small object will make this a herculean feat.

For that reason, the leading approach remains gravitational wave detection. Gravitational waves, "ripples" in space-time that travel at the speed of light, are generated by the acceleration of black holes and other massive bodies. The size of the wave indicates the intensity of gravity as a mass moves through space. Highly energetic events, such as stars exploding or black holes colliding, produce the strongest gravitational waves. The energy of these waves diminishes as they travel through space, offering clues about the distance they've traveled and the size of the objects generating them. First predicted by Einstein in 1916, gravitational waves remained hypothetical until 1974, when astronomers Joseph Taylor Jr. and Russell Hulse inferred their existence from radio waves emitted by a binary pulsar-neutron star system. Binary pulsar-neutron systems consist of two neutron stars, one of which is a pulsar that radiates electromagnetic waves, including radio waves, from its poles. Neutron stars form as massive stars collapse in on themselves, compressing their matter into a core smaller than that of a white dwarf star. In fact, neutron stars are one of the densest celestial objects, second only to black holes. Because of their high mass and proximity to each other, a binary system of neutron stars seemed likely to emit gravitational waves.

After spending four years measuring these signals, Taylor and Hulse realized that the interval of the radio waves they detected was becoming shorter—at a rate of mere microseconds per year. This shift meant the orbit was decaying as the stars experienced gravitational interactions. As this happened, the orbit was losing energy. Einstein had predicted that such energy would produce gravitational waves. When Taylor and Hulse compared the energy loss due to orbital decay with the energy loss predicted by Einstein, the amounts were equivalent. In detecting this binary pulsar-neutron star system, the duo had indirectly confirmed the existence of gravitational waves. In 1993 Hulse and Taylor won the Nobel Prize in Physics for their work on gravitational waves.

The piece de resistance came in 2015, when the Laser Interferometer Gravitational-Wave Observatory (LIGO), a joint effort of the Massachusetts Institute of Technology and the California Institute of Technology, detected gravitational waves caused by the collision of two black holes over a billion light years away. By the time the gravitational waves reached LIGO, they were 10,000 times smaller than the nucleus of an atom. That scientists could detect them is nothing short of remarkable, earning the team of Barry Barish, Kip Thorne, and Rainer Weiss the 2017 Nobel Prize in Physics.

Friction may help scientists as they seek out primordial black holes. Changes in the orbital decay of celestial interactions, caused by dynamical friction, may be visible in gravitational waves, which scientists can now detect. As celestial objects move closer to each other and interact, they will lose energy, emitting gravitational waves with decreasing amplitudes. Soon, we may be able to measure the gravitational waves of nearby celestial objects as they interact with and potentially merge with primordial black holes.[36]

From Dust to Planet

Our solar system formed within the Milky Way from a dense cloud of interstellar dust. Nearly 4.6 billion years ago, energy from a supernova explosion created a cloud of interstellar gas and dust that collapsed in on itself, forming a swirling disk of material. The gravity of the collapse caused the matter to spin faster and faster, pulling in most of the mass.

This created so much pressure in the core that hydrogen atoms combined to form helium, generating large amounts of energy. The temperature continued to rise, and from this condensed, gaseous disk, the protosun formed. Farther out in the disk, the cosmic dust particles were drawn together by gravitational forces and formed clumps of rock. These were planetesimals, the building blocks of planets. Over millions of years, planetesimals collided with each other, combining and growing to form embryonic planets known as protoplanets.

Friction played a central role in the formation of the Milky Way and solar systems in other galaxies, influencing the gravitational interactions between embryonic planets as well as their orbits. We have already seen how tidal friction affects the orbits of moons and their host planets. But that isn't the only mechanism behind planetary migration. The size of planetesimals and the shape of their orbits also depends on dynamical friction. Dynamical friction decelerates large embryos as they move past each other, eventually leading to planetary accretion. As protoplanets grow, they experience more interactions with planetesimals and other protoplanets. The larger they grow, the stronger these interactions become. Depending on how matter is distributed in the developing planetary system, these interactions can pull the protoplanet toward or away from the center of the disk. This is known as planetary migration. Typically, most of the matter in the disk is concentrated near the center, so the developing planet is gravitationally pulled in that direction; however, in some cases, the protoplanet forms a sufficient distance away from the center that it's instead pulled by matter in the opposite direction. As the protoplanet is pulled toward or away from the center, the drag of dynamical friction reduces its energy and speed, influencing the orbital mechanics, just as it does with black hole systems.

The influence of drag persists in planetary systems even after planets form. In 2022 a team from Harvard simulated the effects of drag acting on a gaseous planet. They considered gravitational drag from dynamical friction as well as from hydrodynamic drag, fluid drag caused by the friction of the planet's interacting gases. They found that increased dynamical friction caused a more eccentric orbit while hydrodynamic drag created a more circular orbit. The planet's ultimate orbit was a confluence of both types of friction.[37] Dynamical friction and hydrodynamic drag may not be the only sources of friction influencing a planet's ultimate orbit. Researchers have begun to ask whether dark matter particles form a viscous medium and if this viscosity should be factored into models of the evolution of the universe.[38]

Quantum Friction

As telescopes and measurement techniques evolve to elucidate friction at the largest scales of the universe, labs on Earth are grappling with the opposite challenge: how to detect friction on the quantum scale. Friction must be isolated from other forces acting on the system, making it particularly difficult to measure. Studying the behavior of friction at the quantum scale introduces an entirely new set of challenges that takes our journey with friction into quantum mechanics.

Quantum physics is the field of physics concerned with the fundamental building blocks of matter and energy: atoms and subatomic particles. Scientists have long suspected the existence of quantum friction, or drag, at these levels. When the electrons of fluorescent molecules are excited to a higher level by light, something causes them to return to rest, releasing photons—what we see as fluorescence. Could this suggest the presence of friction at the quantum level? Suspecting quantum friction exists is one thing. Proving its existence is another, necessitating continuous improvements in our ability to detect subatomic interactions.

Even the simplest definition of quantum friction requires peeling back layers of quantum physics. Physicists have theorized that quantum friction is a lateral force experienced by two objects as a result of quantum fluctuations. When we scale down to the quantum level, particles are in constant motion. Energy fluctuates as particles jitter around. We can't know both the exact position and velocity of these quantum particles—once you measure the velocity, the position has already changed, a dilemma illuminated by Werner Heisenberg in his famous uncertainty principle. The question is whether these fluctuations cause friction between two moving objects. To answer such a question, quantum physicists have developed a theory of quantum friction that requires three conditions be met.

First, bodies must be uncharged, or neutral. When there's an imbalance between protons and electrons in an atom, the atom has a charge. We call positively charged atoms *cations* and negatively charged atoms *anions*. Particles become charged when energy is introduced through chemical reactions, causing electrons to be donated or gained. Since ions are unstable, the charged atom will seek out stability by either donating or receiving electrons from neighboring atoms. This exchange of electrons can generate lateral forces that would mask the force caused by quantum friction. To measure this incredibly small quantum friction force, the theory is concerned only with neutral atoms. This is one way to eliminate noise in the system.

Second, the atom must be polarizable. Electrons reside in an electron cloud, the negatively charged region surrounding the atom's nucleus. Polarizability means that the shape of the electron cloud becomes distorted instead of being symmetrical about the nucleus. This causes one side of the atom to end up more positively or negatively charged than the other side and generate van der Waals forces. These weak but long-range forces typically occur in the presence of a charged atom or molecule, or when a neutral atom experiences an electric field created by charged particles. Quantum friction predicts that two polarizable but *neutral* bodies in relative motion will experience a force that opposes that motion. That means neither body in the system is causing an electric field that will induce polarization.

Third, quantum friction must be measured at zero temperature and within a vacuum. This eliminates any possibility of electric fields or thermal activity, reducing noise in the system of interest. Eliminating electric and thermal activity leaves only quantum fluctuations to be measured. Quantum fluctuations are random, temporary changes in energy at a given point in space. The energy in the quantum field, which spans an array of points, will remain constant. But because of the uncertainty principle, points of energy will spontaneously appear and disappear. These quantum fluctuations can cause polarization.

In classical physics, we consider a vacuum environment to be, well, nothing—a void. In quantum physics, however, the vacuum is filled with quantum energy, particles known as *virtual particles* that flit about, instantaneously and temporarily changing the energy at a point in space. These fluctuations are invisible and they're subtle, but they impact electrons, which is how scientists inferred their existence in the first place. For example, two uncharged and perfectly conducting plates held close together in a vacuum shouldn't experience an attractive force, but they do. While the net charge on the system remains zero, atoms in the plates can be attracted to each other due to the distortion of their electron clouds caused by quantum fluctuations. These fluctuations produce drag and are the source of quantum friction. The absence of any other fluctuation or forces in the system means that this drag is, indeed, quantum friction.[39]

There is skepticism around the impact of such small forces that might easily be masked by other interactions. However, these seemingly undetectable forces may cause noticeable changes in the systems around us. Take, for example, an experiment from a couple of decades ago in which researchers flowed water through a membrane made of carbon nanotubes. Carbon nanotubes are hollow cylinders composed of carbon atoms that have diameters on the nanometer scale. The flow of the water defied expectations, running through the carbon nanotubes faster than predicted. In 2022 researchers at the Flatiron Institute in New York City and the French National Center for Scientific Research recreated the experiment, using carbon nanotubes of different radii. In fluid dynamics, as a pipe becomes narrower, water flowing through it slows down due to increased drag against the wall. In the carbon nanotubes, however, water flow sped up as the diameter decreased.[40]

The researchers believe quantum friction is behind this unexpected result. Most work on quantum friction has considered solid interactions—a metal in an electron gas cloud, or even a solid particle in a light field. But in the carbon nanotube experiment, liquid takes the stage. The carbon nanotubes become the smooth, neutral surface in the quantum friction model. Water is now the neutral but polarizable body, with one end of the molecule having a different charge than the other. The electrons shared between the hydrogen and oxygen atoms are not distributed evenly: the oxygen atom pulls more of the electrons toward itself, making it slightly more negative than the hydrogen atoms. This charge difference helps water molecules attract one another. The slightly negative oxygen of one molecule attracts the slightly positive hydrogen of another. Meanwhile, the continuous hexagonal lattice structure of carbon nanotubes is primed for interacting with the water. In carbon nanotubes, the carbons are bonded to three other carbon atoms, enabling each carbon's fourth electron to move readily. When a slightly charged water molecule passes along the wall of a carbon nanotube, the free electron pushes and pulls on it, generating quantum friction.[41] The implications of this could be vast; after all, we're no longer talking about the abstract world of vacuums but water flowing through engineered membranes. Understanding how quantum friction influences flow could inform how we design filtration and desalination systems.

The applications of quantum friction range from the practical—for instance, quantum manipulation of lubricity—to more fantastical ideas, such as extracting energy from a vacuum. These, of course, depend on quantum friction being real. There's been a fair amount of debate and even contention over whether quantum friction exists.[42] But as some point out, there was a time when van der Waals interactions had been predicted but were not universally accepted. It was only decades later that they were measured and proven. Quantum friction is at that in-between stage right now. While researchers have mathematically supported the existence of quantum friction,[43] the real challenge remains experimentally detecting it. The force could be so faint that our technology has yet to detect it. That doesn't mean it won't get there.

The possibility of technological innovation is behind some theorists' perspective that scientists should pursue experiments to confirm quantum friction. While contrarians may argue that a force so faint doesn't particularly matter, the discoveries that lead to its detection may ultimately lead to bigger things, such as controlling the speed of nanoparticles; enabling better precision measurement sensors; or generating magnetic fields in otherwise nonmagnetic materials, enabling us to magnetize mechanical operations without being constrained to a subset of metals. As we continue our quest toward miniaturization, the influence of quantum-level forces may become increasingly significant; friction might once again find itself as a helpful tool for controlling tiny components within MEMS, or it just as easily might be found to be a hindrance. Furthermore, detecting quantum friction will enhance our understanding of quantum fluctuations, pushing the frontiers of quantum physics.[44]

DNA and Protein Friction

In addition to quantum mechanics and astrophysics, tribologists are beginning to turn their tools to investigating sources of friction in the human body. One area where friction may play a key role is in helping us fight infections. Another is in understanding the complex processes by which every protein molecule folds to carry out its precise function.

Bacteriophages, also called phages, are viruses that specifically infect bacteria. The name literally means "to destroy bacteria." Scientists and doctors have been exploring the idea of using phages to treat infections, and in some cases, experimentation has begun. Clinical trials have demonstrated their remarkable efficacy in treating ear infections and leg ulcers. In a world of increasing resistance to bacteria, the promise of such treatment is enormous and exciting. Success hinges on understanding the process of bacteriophage infection, all the way down to how friction influences DNA.

A cell has three main components: the cell membrane, cytoplasm, and nucleus. The cell membrane encases the entire cell, acting as the gatekeeper to the cell through its various protein channels and receptors, which transport crucial substances into and out of the cell. The cell membrane contains the cytoplasm, a water-based environment that houses the machinery of the cell, its organelles. The organelles carry out such functions as assembling proteins and producing the chemical energy that fuels the cell's reactions. At the center of the cell is its nucleus. Technically, the third major component of a cell is its genetic material, or nucleic acids (DNA and RNA), since not every cell has a nucleus. In eukaryotic cells, found in multi-celled organisms like us, DNA and RNA are stored in the nucleus, which is surrounded by a membrane; in prokaryotic cells like bacteria, however, there is no nuclear envelope. Instead, the genetic material of the cell resides in either the inner membrane of the cell or at the transcription site where the RNA copies the DNA gene sequence to create proteins.

Viruses such as phages are not really cells but rather genetic material packaged up by proteins in a protein shell called a capsid. For infection to occur, phage DNA must be injected into the host cell. An infection can either be latent or lytic. A *latent* infection lies dormant, sometimes for the rest of the host's life. It reactivates when the host's immune system is compromised. For example, the virus that causes chicken pox can remain dormant in our body for decades before reactivating as shingles. A *lytic* infection proliferates immediately, using the host cell to make copies of itself before eventually killing the cell. For phages to destroy bacteria, the infection must be lytic.

Some phages are capable of latent and lytic infections. What determines whether an infection is latent or lytic? A lytic infection usually occurs when multiple phages slowly eject their genomes, staggering the timing. If they eject DNA too quickly, the phage integrates into the host cell, and the infection is more likely to be latent. When the DNA doesn't all arrive at once, the slow trickle reduces the cell's resources and leads to a lytic infection. The aim, then, is to kill bacteria by achieving slow DNA ejection of the phage.[45]

Friction can limit or control the speed of DNA ejection by limiting the mobility of DNA molecules. DNA can either be solid or liquid, but when packaged inside a cell, it is usually solid. Fluids usually have lower friction than solids. This is why we add traditional lubricants like greases and water between solid surfaces whenever possible. When DNA is more fluid, rapid ejection occurs across multiple phages simultaneously. If, instead, the DNA is more solid, the ejection process is slower. As DNA transitions from a solid-like state to a more fluid state, friction decreases as a result of the increased spacing between the DNA strands.[46] The separation may allow for hydration between the strands, which would reduce sliding friction, just as providing lubricant between two dry surfaces lowers friction.[47]

In 2014 scientists were examining phages that infect *Escherichia coli,* a common bacterium in the human gut that we mostly hear about when certain strains cause food poisoning.[48] The team gently pressed an AFM tip against the bacterium to measure DNA's stiffness and deformation during ejection at different temperatures. What they found offered remarkable insight into how bacterial viruses have evolved to adapt to hosts: the solid-to-fluid transition of the DNA ejection of the phage was temperature dependent, occurring at around 37 °C, the temperature of the human body. This transition to the fluid state meant that the phage spontaneously released when introduced to the *E. coli* bacterium. Outside the bacterial host, the solid-like DNA meant the structure was stable and the viral genome essentially immobile. Such behavior can be used to tailor treatments, allowing phages that will infect bacteria to be stable when stored on shelves at room temperature and activated only inside the body, where the DNA can eject and attack the targeted bacteria.

In addition to AFM, researchers have employed other techniques to probe the dynamics of DNA ejection, including how viruses eject DNA to infect hosts. One approach involves measuring the energy changes that occur during ejection. This has shown that solid-like DNA requires more energy to be ejected. This is likely due to the higher friction that comes from solid-solid interactions.[49] However, while the dynamics of ejection have been studied extensively, raising awareness of the role of friction in the process, the nature and mechanisms of friction are just beginning to be explored. As neighboring DNA helices from the coiled genome slide over each other, they experience sliding friction forces. In 2023 a team at the University of California San Diego (UCSD) developed a method to more directly probe the role of friction in DNA ejection using optical tweezers. Optical tweezers use lasers to move microscopic objects, grabbing and manipulating them. The team was measuring the dynamics of a phage capsid with tightly packed DNA. Using the optical tweezers, the team pulled the tail of the DNA out of the capsid through a nanochannel in the viral shell, applying a known amount of force. At the same time, they measured the tail's ejection velocity, or how quickly it emerged. As they varied the force and how packed the capsids were, they found a force dependence, related to friction, that coincided with the packing of the DNA microsphere.[50] Friction was higher when the DNA strands were packed tighter and thus rubbed against each other more.

This result aligned with theoretical work showing that friction is also influenced by the shape of DNA—whether it's knotted or neatly coiled.[51] Interestingly, that theoretical work had also predicted that the ejection velocity would be significantly faster than what the UCSD team found.[52] This is because friction is variable during the ejection process. In the UCSD experiment, sliding friction dominated ejection dynamics for the first 20 percent of DNA ejection. The more the microsphere was packed with DNA, the slower the ejection. Higher friction from the interacting DNA strands was the culprit. Because friction clogs can slow down the ejection of DNA, friction can be the difference between a phage successfully infecting a virus or not. With this groundbreaking technique for studying DNA friction, we have a new tool for manipulating and controlling viral infections. For therapeutics, ejection dynamics will ultimately control the effectiveness of the phage against the virus it's targeting. The shape, packing, and arrangement of the DNA in a phage can be designed to tune friction, ensuring that the infection time of the phage allows for the virus to be successfully destroyed. During viral spread, DNA must be ejected from the protein capsid shell. Proteins are chains of amino acids. Some proteins are created in our bodies; others come from our diet. The backbones of proteins are relatively short and simple, made up of carbon, nitrogen, oxygen, and hydrogen. Side chains composed of these elements and sometimes sulfur branch off from the backbone. The genetic information that determines when, in what cells, and in what quantity a protein will be made is stored in DNA and transcribed into

RNA. RNA that is encoded with the information to make a protein is called messenger RNA, or mRNA. It transports the information to the cell's cytoplasm, where an organelle called the ribosome synthesizes the protein. The ribosome does this by reading the mRNA information and then utilizing another molecule, called transfer RNA (tRNA), to assemble the protein, one amino acid at a time. As these amino acid chains are created, some will have a neutral charge, and others will have charged ends. The charge or neutrality of these acids influences their movement and behavior as the final chain of amino acids folds into a three-dimensional structure that determines the function of the protein.

Proteins are the workhorses of our cells. They're involved in almost all the body's functions, including structural support, communication, and transport. Structural proteins enable our bodies to move by providing structural support to cells. Collagen is the primary structural protein in cartilage, discussed in Chapter 4. Meanwhile, messenger proteins send signals throughout the body to coordinate processes in our cells, tissues, and organs. These proteins include hormones such as insulin, which controls glucose levels in our bodies, as well as defense proteins that recognize and respond to pathogens, protecting us from infections. Information is transmitted throughout our brain and from our brain to our nervous system via transport proteins.

This is all to say that proteins do a lot. To function properly, a protein must fold into its most stable configuration, called its native state. Since protein structure determines function, it is essential that a protein adopt the right shape. But folding is not a guaranteed success. When it goes wrong, a protein becomes a nonfunctional clump of material that can be toxic to other cells. Misfolded proteins have gained a lot of attention in recent years for their role in neurological diseases such as Alzheimer's disease and Parkinson's disease. Over six million Americans currently suffer from Alzheimer's and half a million from Parkinson's. It isn't entirely clear why protein folding fails and why misfolded proteins are toxic to other cells. Understanding the mechanisms of folding could lead to cutting-edge therapeutics for diseases involving misfolded proteins.

Unsurprisingly, millions of research dollars are spent each year to investigate protein folding. Advances in AI have been a game changer, enabling researchers to predict the rate of protein folding and dynamics of protein structures as well as what kinds of protein structures will form from different amino acid sequences. Yet many more questions remain, such as why folding sometimes fails. Much like DNA ejection, folding is influenced by a variety of factors, including temperature and the acidity of the cell. Internal friction of the system is also a key contributor to the dynamics of the process and needs to be understood in order to correctly simulate and predict protein folding.

The internal friction of folding is not a straightforward parameter to measure. For this reason, the friction of protein folding is typically measured by time, not force. In lab-controlled experiments of protein folding and unfolding, researchers noticed that protein dynamics were slower than the theoretical dynamics their models predicted.[53] The faster the protein folded, the more pronounced the discrepancy. This puzzled the researchers because they had already accounted for the friction of the protein medium in their model. Even as they repeated their experiment, using media with different viscosities, the process was slower than anticipated. They concluded that the discrepancy was due to internal friction generated by the act of folding.[54]

Determining the causes of internal friction and its influence throughout the folding process is a complex task, not least because of the diversity of proteins and their folding rates, which range from microseconds to milliseconds. Protein folding was once believed to occur in discrete stages of defined structures. Now scientists believe that proteins instead organize into ensemble structures throughout the folding process rather than adopting a few defined structures along the way. The theory behind this is known as the *energy landscape theory* of protein folding. An energy landscape for a protein is a map of the possible energy states of the protein system. The information in the energy landscape helps determine the dynamics and thermodynamics of the system, including the folding process. You can visualize this energy landscape as a surface roughness map, with hills and valleys representing local high- and low-energy barriers. The smoothness or bumpiness of the map affects internal friction. When researchers compared the observed rates of a fast-folding helical protein to the energy landscape, they found that for the protein to rotate and fold around its backbone via covalent bonds between amino acids in the molecule, it had to cross an energy barrier. Doing so caused the internal friction to increase.[55]

This research has raised the question of how much internal friction influences the process of folding versus thermodynamic factors, like the activation energy required for folding. Studying unfolded proteins is one way to isolate the effects and mechanisms of internal friction. It eliminates the complex dynamics and thermodynamics of the folding process. Even the backbone of the protein experiences internal friction. For instance, it takes proteins longer to relax than would be expected based on the effects of friction from interacting with the protein medium. This means that internal friction likely stems from intrachain interactions, such as collisions with other parts of the protein, and rotations about the backbone of the molecule, which require crossing an energy barrier and thus increase friction.[56] Curiously, however, some researchers have found that internal friction is not dependent on chain length. One might assume that with longer chains, there would be more intrachain interactions to generate friction. If internal friction is independent of the chain length, its mechanism is likely of molecular origin, such as hydrogen or ionic bonding. During folding, these bonds form and break between chains, slowing down folding dynamics. However, the relative contribution of internal friction is higher for shorter proteins. This is because longer chains are more exposed to the protein medium, which is itself a source of friction. As the friction of the medium starts to contribute more to the overall friction of folding dynamics, the relative influence of internal friction decreases.[57]

Bridges are another mechanism of internal friction during folding. They form when the positively charged side of an amino acid chain interacts with the negatively charged side of another chain, creating a bridge. During folding, bridges are a significant contributor to internal friction as they hinder movement of the protein. Once the protein is folded into its three-dimensional structure, atoms can be tightly packed, preventing the protein medium from lubricating the contacting atoms. Now the friction stems from the atoms rubbing against each other and interacting as they sit neighboring each other in their folded configuration. These different mechanisms of friction throughout the folding process ultimately contribute to the progress and success of the protein reaching its native state.[58]

With increasingly sensitive experimental techniques and improved computing power for simulations, we're able to better investigate folding dynamics and the factors that influence them. Researchers are beginning to incorporate aspects of friction into their experiments and simulations to understand its role in successful protein folding. Does friction slow down the process of folding so that the protein becomes trapped in an energy well in the energy landscape, unable to adopt its native state? Can friction cause protein folding to fail? Is it possible to manipulate internal friction through temperature, external forces, or other environmental factors or use it to control the dynamics of the process, ensuring success? And can we grasp the mechanisms of friction well enough to predict the success and failure of protein folding and the structures that will form?

The answers to these questions will aid in the development of therapeutics for some of the most challenging and frustrating diseases we face today. Protein-protein interactions (PPIs), aptly named as they describe interactions between two or more proteins, are responsible for a variety of necessary biological functions, such as growth and transporting signals across our bodies. However, some PPIs can also be responsible for diseases, including cancers and neurodegenerative diseases. Therapeutics to modulate and inhibit these undesirable interactions are a targeted area of research. A primary challenge is that developing therapeutics for PPIs requires a thorough understanding of those structures. Monoclonal antibodies (mAbs), for example, use engineered proteins to seek out and attach to a targeted protein, such as Covid-19 or cancer. Antibodies are proteins that the immune system uses to identify and attack viruses, bacteria, and other foreign substances known as antigens. Designed to target a foreign material in the body, mAbs are copies of a specific antibody. The injected mAbs seek out the antigen they're designed for and then attach to it. After the mAbs bind to the antigen, the mAbs can either neutralize it, inhibiting it from replicating and growing, or produce lipids and proteins that break the antigen down. For instance, mAbs are being used to target HER2, a protein behind aggressive, fast-spreading breast cancer. But to design the antibody, we need a good understanding of the protein structure, or the failed protein structure responsible for the disease. This research was once limited by its exorbitant costs. But the Covid-19 pandemic helped invigorate research funding in this area. Additionally, groundbreaking developments in artificial intelligence (AI) have entered the scene, offering the ability to predict the 3D folding structures of proteins. One such program, AlphaFold, is considered one of the most important AI achievements to date, giving anyone with a computer access to its structure-predicting algorithm. The hope is that the costs of PPI therapeutics research will decrease, further accelerating the development of new treatments.

We find ourselves on the cusp of groundbreaking work involving friction in medicine. It is astonishing to think that friction—an invisible force—may stand in the way of the creation of therapeutics for neurological diseases or cancer. But as we've seen throughout this book, friction is omnipresent. It's up to us to understand it. Throughout history, we've probed and expanded our understanding of friction, turning it into a powerful tool. This chapter has brought us to the frontiers of friction, where many theories are still being investigated. What technologies and advances will emerge once we have more clarity on quantum friction and dark matter? Or the friction contributing to the folding of proteins within our bodies? Friction has enabled the wheel and control of fire, inventions that were crucial to civilization as we know it. Now it may help us probe the origins of the universe itself and even our own evolution.

The force that opposes motion is continually driving us forward.

Notes
Acknowledgments
Index

Notes

1 · Tribology Is Born

1. Peter H. Jost, "Tribology: How a Word Was Coined 40 Years Ago," *ARCHIVE Proceedings of the Institution of Mechanical Engineers Part J Journal of Engineering Tribology 1994–1996 (Vols 208–210)* 223 (2009): 240–244.

2. Theo Mang, Kirsten Bobzin, and Thorsten Bartels, *Industrial Tribology: Tribosystems, Friction, Wear and Surface Engineering, Lubrication* (Weinheim: Wiley-VCH, 2011); Peter H. Jost, "Economic Impact of Tribology," in *NBS Special Publication* (Gaithersburg, MD: National Bureau of Standards, 1976).

3. Jost, "Tribology: How a Word Was Coined."

4. Harry van Leeuwen, "Petrus van Musschenbroek (1692–1761), Man of Tribology," *Proceedings of the Institution of Mechanical Engineers, Part J: Journal of Engineering Tribology* 235, no. 12 (2021): 2537–51, https://doi.org/10.1177/13506501211042704.

5. Jost, "Tribology: How a Word Was Coined."

6. Francesco Berna et al., "Microstratigraphic Evidence of In Situ Fire in the Acheulean Strata of Wonderwerk Cave, Northern Cape Province, South Africa," *Proceedings of the National Academy of Sciences* 109, no. 20 (2012): E1215–20, https://doi.org/10.1073/pnas.1117620109.

7. Richard Wrangham, *Catching Fire: How Cooking Made Us Human* (New York: Basic Books, 2009).

8. J. A. J. Gowlett, "The Discovery of Fire by Humans: A Long and Convoluted Process," *Philosophical Transactions of the Royal Society B: Biological Sciences* 371, no. 1696 (2016): 20150164, https://doi.org/10.1098/rstb.2015.0164.

9. Sayed Hemeda and Alghreeb Sonbol, "Sustainability Problems of the Giza Pyramids," *Heritage Science* 8, no. 1 (2020): 8, https://doi.org/10.1186/s40494-020-0356-9.

10. A. Fall et al., "Sliding Friction on Wet and Dry Sand," *Physical Review Letters* 112, no. 17 (2014): 175502, https://doi.org/10.1103/PhysRevLett.112.175502.

11. George Ripley and Charles A. Dana, *The American Cyclopaedia*, vol. 5 (New York: D. Appleton, 1873).

12. Percy E. Newberry, *El Bersheh*, part 1, *The Tomb of Tehuti-Hetep* (London: Offices of the Egypt Exploration Fund, 1895), plate 15, Inner Chamber, left hand wall, upper row, available through Heidelberg Historic Literature—Digitized, Universitätsbibliothek Heidelberg, https://doi.org/10.11588/diglit.4191#0066.

13. Megan Gambino, "A Salute to the Wheel," *Smithsonian*, June 17, 2009, https://www.smithsonianmag.com/science-nature/a-salute-to-the-wheel-31805121/.

14. H. A. Harris, "Lubrication in Antiquity," *Greece & Rome* 21, no. 1 (1974): 32–36; A. H. McDonald, "Cato (1)," in *Oxford Classical Dictionary* (Oxford: Clarendon Press, 1970).

15. Pliny the Elder, *The Natural History,* trans. John Bostock and H.T. Riley (London: Henry G. Bohn, 1895), vol. 5, Book XXVIII, Chapter 37, 324.

16. Harris, "Lubrication in Antiquity."

17. Harris, "Lubrication in Antiquity."

18. Harris, "Lubrication in Antiquity."

19. O. Crumlin-Pedersen and O. Olsen, *The Skuldelev Ships, I. Topography, Archaeology, History, Conservation and Display* (Roskilde, Denmark: Viking Ship Museum, 2002); "The Five Viking Ships," in Exhibitions, Viking Ship Museum, Roskilde, Denmark, n.d., https://www.vikingeskibsmuseet.dk/en/visit-the-museum/exhibitions/the-five-viking -ships.

20. Virginia Mills, "Perpetual Motion," History of Science blog, The Royal Society, September 25, 2018, https://royalsociety.org/blog/2018/09/perpetual-motion/#:~:text =Attempts%20to%20create%20a%20perpetual,forcing%20the%20axle%20to%20 continue.

21. Walter Isaacson, *Leonardo da Vinci* (New York: Simon and Schuster, 2017).

22. Aristotle, *Physics,* Book VIII, Part IV in *The Complete Works of Aristotle,* ed. Jonathan Barnes, vol. 1 (Princeton, NJ: Princeton University Press, 1991).

23. Allan Franklin, "Mechanics, Aristotelian," in *Routledge Encyclopedia of Philosophy* (Taylor and Francis,1998), https://doi.org/10.4324/9780415249126-Q067-1; Duncan Dowson, "Men of Tribology: Leonardo da Vinci (1452–1519)," *Journal of Lubrication Technology* 99, no. 4 (1977): 382–386, https://doi.org/10.1115/1.3453230.

24. Ian M. Hutchings, "Leonardo da Vinci's Studies of Friction," *Wear* 360–361 (2016): 51–66, https://doi.org/10.1016/j.wear.2016.04.019.

25. Dowson, "Men of Tribology: Leonardo da Vinci"; Hutchings, "Leonardo da Vinci's Studies of Friction."

26. Hutchings, Leonardo da Vinci's Studies of Friction"; Dowson, "Men of Tribology: Leonardo da Vinci."

27. Duncan Dowson, "Men of Tribology: Guillaume Amontons (1663–1705) and John Theophilus Desaguliers (1683–1744)," *Journal of Lubrication Technology* 100, no. 1 (1978): 2–5, https://doi.org/10.1115/1.3453109.

28. Guillaume Amontons, "Moyen de substituer commodement l'action du Feu, a la force des Hommes et des chevaux pour mouvoir les machines" (Method of Substituting the Force of Fire for Horse and Man Power to Move Machines), *Mémoires de l'Académie Royale des Sciences,* June 20, 1699, 112–126.

29. Dowson, "Men of Tribology: Guillaume Amontons and John Theophilus Desaguliers"; Marcin Wołowicz, Piotr Kolasiński, and Krzysztof Badyda, "Modern Small and Microcogeneration Systems—A Review," *Energies* 14, no. 3 (2021): 1–47.

30. Jaime Wisniak, "Guillaume Amontons," *Revista CENIC. Ciencias Quimicas* 36, no. 3 (2005): 187–195.

31. Ian M. Hutchings, "A Note on Guillaume Amontons and the Laws of Friction," *Proceedings of the Institution of Mechanical Engineers, Part J: Journal of Engineering Tribology* 235, no. 12 (2021): 2530–36, https://doi.org/10.1177/13506501211039385.

32. Hutchings, "A Note on Guillaume Amontons."

33. Ian M. Hutchings and Philip Shipway, *Tribology: Friction and Wear of Engineering Materials*, 2nd ed. (Elsevier Science, 2017).

34. Richard Williams, "June 1785: Coulomb Measures the Electric Force," APS News: This Month in Physics History, June 2016, http://www.aps.org/publications/apsnews /201606/physicshistory.cfm.

35. Duncan Dowson, "Men of Tribology: Charles Augustin Coulomb (1736–1806) and Arthur-Jules Morin (1795–1880)," *Journal of Lubrication Technology* 100, no. 2 (1978): 148–155, https://doi.org/10.1115/1.3453126.

36. C. A. Coulomb, *Théorie des machines simples*, nouvelle édition (Paris: Bachelier, 1821), https://gallica.bnf.fr/ark:/12148/bpt6k1095299; Yannick Desplanques, "Amontons-Coulomb Friction Laws, A Review of the Original Manuscript," *SAE International Journal of Materials and Manufacturing* 8, no. 1 (2015): 98–103.

37. Dowson, "Men of Tribology: Charles Augustin Coulomb and Arthur-Jules Morin."

2 · Friction: Friend or Foe?

1. Albert Van Helden, *The Invention of the Telescope* (Philadelphia: American Philosophical Society, 1977), http://archive.org/details/inventionofteles0000vanh.

2. Robert M. Hazen, "The Discovery of Gravity and Laws of Motion by Isaac Newton," Wondrium Daily, March 19, 2021; "Galileo Galilei (1564–1642)," *British Journal of Sports Medicine* 40, no. 9 (2006): 806–807; Isaac Newton, *The Mathematical Principles of Natural Philosophy*, trans. Andrew Motte, 2 vols. (London: Printed for B. Motte, 1729), http:// archive.org/details/bub_gb_Tm0FAAAAQAAJ.

3. Newton, *The Mathematical Principles of Natural Philosophy*.

4. Yoel Eli Stuart et al., "Rapid Evolution of a Native Species Following Invasion by a Congener," *Science* 346, no. 6208 (2014): 463–466; Samantha Murray, "UF Study: Where Brown Anoles Invade, Native Green Anoles Reach New Heights," news, Institute of Food and Agricultural Studies, University of Florida, October 7, 2021, https://blogs.ifas.ufl .edu/news/2021/10/07/uf-study-where-brown-anoles-invade-native-green-anoles -reach-new-heights/; Jesse B. Borden, Stephanie Bohlman, and Brett R. Scheffers, "Niche Lability Mitigates the Impact of Invasion but Not Urbanization," *Oecologia* 198 (2022): 1–10; "Florida Lizards Are Evolving, Fast," *Science Connected Magazine* (blog), October 24, 2014, https://magazine.scienceconnected.org/2014/10/florida-lizards -evolving-rapidly/.

5. Heinrich Hertz, "On the Contact of Elastic Solids" (1881), in Hertz, *Miscellaneous Papers by Heinrich Hertz*, trans. D. E. Jones and G. A. Schott (London: MacMillan and Co., 1896).

6. Hertz, "On the Contact of Elastic Solids"; Kenneth L. Johnson, "One Hundred Years of Hertz Contact," *Proceedings of the Institution of Mechanical Engineers* 196, no. 1 (1982): 363–378.

7. "Dr Jim Greenwood Receives the Prestigious Tribology Gold Medal," Department of Engineering, University of Cambridge, November 1, 2021, http://www.eng.cam.ac.uk /news/dr-jim-greenwood-receives-prestigious-tribology-gold-medal.

8. J. A. Greenwood, J. B. P. Williamson, and Frank Philip Bowden, "Contact of Nominally Flat Surfaces," *Proceedings of the Royal Society of London. Series A. Mathematical and Physical Sciences* 295, no. 1442 (1966): 300–319, https://doi.org/10.1098/rspa.1966.0242.

9. Jianping Gao et al., "Frictional Forces and Amontons' Law: From the Molecular to the Macroscopic Scale," *Journal of Physical Chemistry B* 108, no. 11 (2004): 3410–25, https://doi.org/10.1021/jp036362l.

10. Steffen Ducheyne, "The Times and Life of John Th. Desaguliers (1683–1744): Newtonian and Freemason," *Revue Belge de Philologie et d'Histoire* 87, no. 2 (2009): 349–363; Duncan Dowson, "Men of Tribology: Guillaume Amontons (1663–1705) and John Theophilus Desaguliers (1683–1744)," *Journal of Lubrication Technology* 100, no. 1 (1978): 2–5, https://doi.org/10.1115/1.3453109.

11. Ducheyne, "The Times and Life of John Th. Desaguliers"; Dowson, "Men of Tribology: Guillaume Amontons and John Theophilus Desaguliers"; Andrew M. A. Morris, "Evaluating John Theophilus Desaguliers' Newtonianism: The Case of Waterwheel Knowledge in *A Course of Experimental Philosophy*," *Notes and Records: The Royal Society Journal of the History of Science* 74, no. 3 (2019): 453–477, https://doi.org/10.1098/rsnr.2019.0023.

12. Ducheyne, "The Times and Life of John Th. Desaguliers"; Morris, "Evaluating John Theophilus Desaguliers' Newtonianism."

13. Larry Stewart, "'The King George III Collection' at the Science Museum," *Technology and Culture* 35, no. 4 (1994): 857–863, https://doi.org/10.2307/3106509; *J.T Desaguliers—Plate 31 from A Course of Experimental Philosophy (London, 1734–44)*, Image Archive, accessed September 7, 2023, https://calisphere.org/item/d0e8ce73902e0d39c38482dcc59bb68f/; J. T. Desaguliers, *A Course of Experimental Philosophy*, vol. 1 (London, 1734; reprint Kessinger, 2010).

14. Desaguliers, *A Course of Experimental Philosophy*, vol. 1.

15. Ian M. Hutchings, "A Note on Guillaume Amontons and the Laws of Friction," *Proceedings of the Institution of Mechanical Engineers, Part J: Journal of Engineering Tribology* 235, no. 12 (2021): 2530–36, https://doi.org/10.1177/13506501211039385; Yannick Desplanques, "Amontons-Coulomb Friction Laws, A Review of the Original Manuscript," *SAE International Journal of Materials and Manufacturing* 8, no. 1 (2015): 98–103; Martin Trapp and Fang Chen, *Automotive Buzz, Squeak and Rattle: Mechanisms, Analysis, Evaluation and Prevention* (Elsevier, 2011).

16. David Tabor, "Frank Philip Bowden, 1903–1968," *Biographical Memoirs of Fellows of the Royal Society* 15 (1997): 1–38, https://doi.org/10.1098/rsbm.1969.0001.

17. F. P. Bowden and D. Tabor, "Mechanism of Metallic Friction," *Nature* 150, no. 3798 (1942): 197–199, https://doi.org/10.1038/150197a0; F. P. Bowden, A. J. W. Moore, and D. Tabor, "The Ploughing and Adhesion of Sliding Metals," *Journal of Applied Physics* 14, no. 2 (1943): 80–91, https://doi.org/10.1063/1.1714954; F. P. Bowden and G. W. Rowe, "The Adhesion of Clean Metals," *Proceedings of the Royal Society of London. Series A. Mathematical and Physical Sciences* 233, no. 1195 (1956): 429–442, https://doi.org/10.1098/rspa.1956.0001.

18. Ben Shouse, "How Geckos Stick on Der Waals," August 27, 2002, https://www.science.org/content/article/how-geckos-stick-der-waals; Brian J. Briscoe, "Chapter 2—Interfacial Friction of Polymer Composites. General Fundamental Principles," in *Composite Materials Series*, ed. Klaus Friedrich, vol. 1 (Elsevier, 1986), 25–59, https://doi.org/10.1016/B978-0-444-42524-9.50006-5; Bharat Bhushan, *Introduction to Tribology* (John Wiley & Sons, 2013); Ian M. Hutchings and Philip Shipway, *Tribology: Friction and Wear of Engineering Materials*, 2nd ed. (Elsevier Science, 2017).

19. Hutchings and Shipway, *Tribology*.

20. Bhushan, *Introduction to Tribology*.

21. Kenneth L. Johnson, *Contact Mechanics* (Cambridge University Press, 1987); Roderick A. Smith, "Kenneth Langstreth Johnson. 19 March 1925—21 September 2015," *Biographical Memoirs of Fellows of the Royal Society* 62 (2016): 247–265, https://doi.org/10.1098/rsbm.2016.0012.

22. Smith, "Kenneth Langstreth Johnson."

23. Robert W. Carpick, D. Frank Ogletree, and Miquel Salmeron, "A General Equation for Fitting Contact Area and Friction vs Load Measurements," *Journal of Colloid and Interface Science* 211, no. 2 (1999): 395–400, https://doi.org/10.1006/jcis.1998.6027.

24. B. N. J. Persson, "Contact Mechanics for Randomly Rough Surfaces," *Surface Science Reports* 61, no. 4 (2006): 201–227, https://doi.org/10.1016/j.surfrep.2006.04.001.

25. Kenneth L. Johnson, Kevin Kendall, and A. D. Roberts, "Surface Energy and the Contact of Elastic Solids," *Proceedings of the Royal Society of London. A. Mathematical and Physical Sciences* 324, no. 1558 (1971): 301–313, https://doi.org/10.1098/rspa.1971.0141; Robert I. Taylor, "Rough Surface Contact Modelling—A Review," *Lubricants* 10, no. 5 (2022), https://doi.org/10.3390/lubricants10050098.

26. Eduard Arzt, Stanislav Gorb, and Ralph Spolenak, "From Micro to Nano Contacts in Biological Attachment Devices," *Proceedings of the National Academy of Sciences* 100, no. 19 (2003): 10603–6, https://doi.org/10.1073/pnas.1534701100; Austin M. Garner et al., "Going Out on a Limb: How Investigation of the Anoline Adhesive System Can Enhance Our Understanding of Fibrillar Adhesion," *Integrative and Comparative Biology* 59, no. 1 (2019): 61–69, https://doi.org/10.1093/icb/icz012.

27. Austin M. Garner et al., "The Same but Different: Setal Arrays of Anoles and Geckos Indicate Alternative Approaches to Achieving Similar Adhesive Effectiveness," *Journal of Anatomy* 238, no. 5 (2021): 1143–55, https://doi.org/10.1111/joa.13377.

28. Bhushan, *Introduction to Tribology*; Hutchings and Shipway, *Tribology*.

29. A. R. Savkoor, "Models of Friction," section 8.3, in *Handbook of Materials Behavior Models*, ed. Jean Lemaitre (Burlington: Academic Press, 2001), 700–759, https://doi.org/10.1016/B978-012443341-0/50075-2; Bhushan, *Introduction to Tribology*.

30. Bhushan, *Introduction to Tribology*.

31. "Earthquakes Can Make Thrust Faults Open Violently and Snap Shut: Experiments Reveal a New Mechanism That Could Explain the Source of a Destructive Feature of the 2011 Tohoku Earthquake," ScienceDaily, May 1, 2017, https://www.sciencedaily.com/releases/2017/05/170501131647.htm.

32. Vahe Gabuchian et al., "Experimental Evidence That Thrust Earthquake Ruptures Might Open Faults," *Nature* 545, no. 7654 (2017): 336–339, https://doi.org/10.1038/nature22045.

33. Santiago Casado, "Studying Friction While Playing the Violin: Exploring the Stick–Slip Phenomenon," *Beilstein Journal of Nanotechnology* 8 (2017): 159–166, https://doi.org/10.3762/bjnano.8.16.

34. Bhushan, *Introduction to Tribology;* Hutchings and Shipway, *Tribology.*

35. Bhushan, *Introduction to Tribology;* Hutchings and Shipway, *Tribology.*

36. Hutchings and Shipway, *Tribology.*

37. Yun Dong et al., "Phononic Origin of Structural Lubrication," *Friction* 11, no. 6 (2023): 966–976, https://doi.org/10.1007/s40544-022-0636-3.

38. Nikolay T. Garabedian, "A Direct Experimental Link between Atomic-Scale and Macroscale Friction" (PhD diss., University of Delaware, 2019), http://udspace.udel.edu/handle/19716/24857; Jeffrey L. Streator, "Nanoscale Friction: Phonon Contributions for Single and Multiple Contacts," *Frontiers in Mechanical Engineering* 5 (2019), https://www.frontiersin.org/articles/10.3389/fmech.2019.00023.

3 · When You Rub the Wrong Way

1. Paul Cooper, "Nemi Ships: How Caligula's Floating Pleasure Palaces Were Found and Lost Again," *Discover,* November 7, 2018, https://www.discovermagazine.com/planet-earth/nemi-ships-how-caligulas-floating-pleasure-palaces-were-found-and-lost-again; Duncan Dowson and Bernard J. Hamrock, "History of Ball Bearings," NASA Technical Memorandum 81629, National Aeronautics and Space Administration, February 1981, https://ntrs.nasa.gov/api/citations/19810009866/downloads/19810009866.pdf.

2. Marco Ceccarelli et al., "Ball Bearings from Roman Imperial Ships of Nemilake," *Advances in Historical Studies* 8, no. 3 (2019): 115–130.

3. Ian M. Hutchings, "Leonardo da Vinci's Studies of Friction," *Wear* 360–361 (2016): 51–66, https://doi.org/10.1016/j.wear.2016.04.019.

4. Dowson and Hamrock, "History of Ball Bearings."

5. Dowson and Hamrock, "History of Ball Bearings."

6. Michael S. Craig, Crayons containing ester waxes, World Intellectual Property Organization WO1999065997A1, filed June 17, 1999, and issued December 23, 1999, https://patents.google.com/patent/WO1999065997A1/en.

7. Kian Kun Yap et al., "Wax-Oil Lubricants to Reduce the Shear between Skin and PPE," *Scientific Reports* 11, no. 1 (2021): 11537, https://doi.org/10.1038/s41598-021-91119-0.

8. Bharat Bhushan, *Introduction to Tribology* (John Wiley & Sons, 2013); Ian M. Hutchings and Philip Shipway, *Tribology: Friction and Wear of Engineering Materials,* 2nd ed. (Elsevier Science, 2017).

9. John Emsley, *Nature's Building Blocks: An A-Z Guide to the Elements* (Oxford University Press, 2011).

10. Agency for Toxic Substances and Disease Registry, "Chemical and Physical Information," ch. 4 in *Toxicological Profile for Molybdenum* (Atlanta: National Library of Medicine, 2020), https://www.ncbi.nlm.nih.gov/books/NBK590373.

11. European Association of Geochemistry, "We May All Be Martians: New Research Supports Theory That Life Started on Mars," August 28, 2013, https://phys.org/news/2013-08-martians-theory-life-mars.html.

12. Geoff Rayner-Canham and Jonathan Grandy, "Molybdenum and Evolution," RSC Education, Royal Society of Chemistry, August 31, 2011, https://edu.rsc.org/feature/molybdenum-and-evolution/2020196.article; Lucas Brouwers, "A Spoonful of Molybdenum, Some Ulysses and the Origin of Life," Scientific American, *Thoughtomics* (blog), April 12, 2012, https://blogs.scientificamerican.com/thoughtomics/a-spoonful-of-molybdenum-some-ulysses-and-the-origin-of-life/.

13. Kazuhisa Miyoshi, "Solid Lubrication Fundamentals and Applications: Introduction and Background," NASA Technical Memorandum 107249, National Aeronautics and Space Administration, July 1998, https://ntrs.nasa.gov/api/citations/19980218923/downloads/19980218923.pdf.

14. "This Month in Physics History April 6, 1938: Discovery of Teflon," *APS News* 30, no. 4 (April 2021), http://www.aps.org/publications/apsnews/202104/history.cfm.

15. W. Thomas Chase, "Egyptian Blue as a Pigment and Ceramic Material," in *Science and Archaeology,* ed. Robert H. Brill (Cambridge, MA: MIT Press, 1971), 80–90.

16. R. E. Jones and E. Photos-Jones, "Technical Studies of Aegean Bronze Age Wall Painting: Methods, Results and Future Prospects," *British School at Athens Studies* 13 (2005): 199–228.

17. David L. Chandler, "Explained: Chemical Vapor Deposition," MIT News, Massachusetts Institute of Technology, June 19, 2015, https://news.mit.edu/2015/explained-chemical-vapor-deposition-0619.

18. Al Powell, "Defending Breakthrough Research," *Harvard Gazette,* June 24, 2016, https://news.harvard.edu/gazette/story/2016/06/defending-breakthrough-research/

19. Abdullah Saleh Algamili et al., "A Review of Actuation and Sensing Mechanisms in MEMS-Based Sensor Devices," *Nanoscale Research Letters* 16, no. 1 (2021): 16, https://doi.org/10.1186/s11671-021-03481-7.

20. C. Mathew Mate and Robert W. Carpick, *Tribology on the Small Scale* (Oxford: Oxford Graduate Texts, 2019).

21. Mate and Carpick. *Tribology on the Small Scale.*

22. Mate and Carpick. *Tribology on the Small Scale.*

23. D. W. Wang et al., "How Do Grooves on Friction Interface Affect Tribological and Vibration and Squeal Noise Performance," *Tribology International* 109 (2017): 192–205, https://doi.org/10.1016/j.triboint.2016.12.043.

24. "Anti-Friction Coatings Market Size, Opportunities by 2028," accessed September 5, 2023, https://www.databridgemarketresearch.com/reports/global-anti-friction-coating-market.

25. Judy Runge, "A Brief History of Anodizing Aluminum," in Runge, *The Metallurgy of Anodizing Aluminum* (Cham: Springer, 2018), ch. 1.

26. Andreas Almqvist et al., "A Scientific Perspective on Reducing Ski-Snow Friction to Improve Performance in Olympic Cross-Country Skiing, the Biathlon and Nordic Combined," *Frontiers in Sports and Active Living* 4 (2022), https://www.frontiersin.org/articles/10.3389/fspor.2022.844883.

4 · Going Against the Flow

1. Bharat Bhushan, *Introduction to Tribology* (John Wiley & Sons, 2013); E. N. Da C. Andrade, "Newton and the Science of His Age," *Proceedings of the Royal Society of London. Series B, Biological Sciences* 131, no. 864 (1943): 207–223.

2. M. le docteur (Jean-Léonard-Marie) Poiseuille and Royal College of Surgeons of England, *Recherches sur la force du coeur aortique* (Paris: De l'impr. de Didot le jeune, 1828), http://archive.org/details/b22291611.

3. J. Pfitzner, "Poiseuille and His Law," *Anaesthesia* 31, no. 2 (1976): 273–275, https://doi .org/10.1111/j.1365-2044.1976.tb11804.x.

4. S. P. Sutera and R. Skalak, "The History of Poiseuille's Law," *Annual Review of Fluid Mechanics* 25, no. 1 (1993): 1–20, https://doi.org/10.1146/annurev.fl.25.010193.000245.

5. Sutera and Skalak, "The History of Poiseuille's Law."

6. Sutera and Skalak, "The History of Poiseuille's Law."

7. Bhushan, *Introduction to Tribology;* Ashlie Martini, *Introduction to Tribology for Engineers* (independently published, 2022).

8. Sudha Balakrishnan and Anthony B. Ward, "The Diagnosis and Management of Adults with Spasticity," in *Handbook of Clinical Neurology,* ed. Michael P. Barnes and David C. Good, vol. 110 (Elsevier, 2013), 145–160, https://doi.org/10.1016/B978-0-444-52901 -5.00013-7.

9. Duncan Dowson, "Men of Tribology: Robert Henry Thurston (1839–1903) and Osborne Reynolds (1842–1919)," *Journal of Lubrication Technology* 100, no. 4 (1978): 455–461, https://doi.org/10.1115/1.3453250; "This Month in Physics History: March 15, 1883: Osborne Reynolds Proposes the Reynolds Number," *APS News,* March 2020, http://www.aps.org/publications/apsnews/202003/history.cfm.

10. "This Month in Physics History: March 15, 1883: Osborne Reynolds Proposes the Reynolds Number."

11. Osborne Reynolds, "XXIX. An Experimental Investigation of the Circumstances Which Determine Whether the Motion of Water Shall Be Direct or Sinuous, and of the Law of Resistance in Parallel Channels," *Philosophical Transactions of the Royal Society of London* 174 (1883): 935–982, https://doi.org/10.1098/rstl.1883.0029; Dowson, "Men of Tribology: Robert Henry Thurston and Osborne Reynolds."

12. Osborne Reynolds, "IV. On the Theory of Lubrication and Its Application to Mr. Beauchamp Tower's Experiments, Including an Experimental Determination of the Viscosity of Olive Oil," *Philosophical Transactions of the Royal Society of London* 177 (1886): 157–234, https://doi.org/10.1098/rstl.1886.0005.

13. Dowson, "Men of Tribology: Robert Henry Thurston and Osborne Reynolds"; Reynolds, "IV. On the Theory of Lubrication."

14. Dowson, "Men of Tribology: Robert Henry Thurston and Osborne Reynolds"; Debbie Sniderman, "Robert Henry Thurston," American Society of Mechanical Engineers, June 22, 2012, https://www.asme.org/topics-resources/content/Robert-Henry-Thurston.

15. Thomas Christian Thomsen, *The Practice of Lubrication: An Engineering Treatise on the Origin, Nature and Testing of Lubicants, Their Selection, Application and Use* (McGraw-Hill, 1920); Dowson, "Men of Tribology: Robert Henry Thurston and Osborne Reynolds."

16. R. H. Thurston, Machine for Testing Lubricants, 230158, issued July 20, 1880; Thomsen, *The Practice of Lubrication.*

17. Mathias Woydt and Rolf Wäsche, "The History of the Stribeck Curve and Ball Bearing Steels: The Role of Adolf Martens," *Wear* 268, no. 11 (2010): 1542–46, https://doi.org /10.1016/j.wear.2010.02.015.

18. Duncan Dowson, "Men of Tribology: Heinrich Rudolph Hertz (1857–1894) and Richard Stribeck (1861–1950)," *Journal of Lubrication Technology* 101, no. 2 (1979): 115–119, https://doi.org/10.1115/1.3453287; R. Stribeck, *Die wesentlichen Eigenschaften der Gleit- und Rollenlager* (Berlin: Springer, 1903).

19. Martini, *Introduction to Tribology for Engineers.*

20. Martini, *Introduction to Tribology for Engineers.*

21. Alexander H. Tullo, "Engine Oil Becomes Critical as Automakers Look to Boost Gas Mileage," Chemical & Engineering News, American Chemical Society, February 3, 2019, https://cen.acs.org/business/specialty-chemicals/Engine-oil-becomes-critical -automakers/97/i5.

22. James A. Greenwood, "Elastohydrodynamic Lubrication," *Lubricants* 8, no. 5 (2020), https://doi.org/10.3390/lubricants8050051.

23. Kenneth L. Johnson, J.A. Greenwood, and S.Y. Poon, "A Simple Theory of Asperity Contact in Elastohydro-Dynamic Lubrication," *Wear* 19, no. 1 (1972): 91–108, https://doi .org/10.1016/0043-1648(72)90445-0.

24. Katharine B. Blodgett, "Films Built by Depositing Successive Monomolecular Layers on a Solid Surface," *Journal of the American Chemical Society* 57, no. 6 (1935): 1007–22, https://doi.org/10.1021/ja01309a011.

25. H. A. Spikes and P. M. Cann, "The Development and Application of the Spacer Layer Imaging Method for Measuring Lubricant Film Thickness," *Proceedings of the Institution of Mechanical Engineers, Part J: Journal of Engineering Tribology* 215, no. 3 (2001): 261–277, https://doi.org/10.1243/1350650011543529; "Katharine Burr Blodgett," *Physics Today* 33, no. 3 (1980): 107, https://doi.org/10.1063/1.2913969.

26. M. Whelan and Edwin Reilly, Jr., "Katharine Burr Blodgett," Engineering Hall of Fame, Edison Tech Center, n.d., https://edisontechcenter.org/Blodgett.html.

27. Spikes and Cann, "The Development and Application of the Spacer Layer Imaging Method."

28. Spikes and Cann, "The Development and Application of the Spacer Layer Imaging Method."

29. S. Whitehouse et al., "Fluorescent Imaging of Razor Cartridge / Skin Lubrication," *Surface Topography: Metrology and Properties* 9, no. 2 (2021): 024001, https://doi.org/10 .1088/2051-672X/ac0ba2.

30. R. S. Dwyer-Joyce, T. Reddyhoff, and J. Zhu, "Ultrasonic Measurement for Film Thickness and Solid Contact in Elastohydrodynamic Lubrication," *Journal of Tribology* 133, no. 031501 (2011), https://doi.org/10.1115/1.4004105.

31. A. C. Moore and D. L. Burris, "Tribological Rehydration of Cartilage and Its Potential Role in Preserving Joint Health," *Osteoarthritis and Cartilage* 25, no. 1 (2017): 99–107, https://doi.org/10.1016/j.joca.2016.09.018.

214 · NOTES TO PAGES 106–114

32. Jarrett M. Link et al., "The Tribology of Cartilage: Mechanisms, Experimental Techniques, and Relevance to Translational Tissue Engineering," *Clinical Biomechanics* 79(2020): 104880, https://doi.org/10.1016/j.clinbiomech.2019.10.016.
33. Carmine Putignano et al., "Cartilage Rehydration: The Sliding-Induced Hydrodynamic Triggering Mechanism," *Acta Biomaterialia* 125 (2021): 90–99, https://doi.org/10.1016/j.actbio.2021.02.040; Moore and Burris, "Tribological Rehydration of Cartilage."
34. Alison C. Dunn et al., "Lubrication Regimes in Contact Lens Wear during a Blink," *International Conference on BioTribology 2011* 63 (2013): 45–50, https://doi.org/10.1016/j.triboint.2013.01.008; Dong Qin et al., "Tribological Behaviour of Two Kinds of Typical Hydrogel Contact Lenses in Different Lubricants," *Biosurface and Biotribology* 5, no. 4 (2019): 110–117, https://doi.org/10.1049/bsbt.2019.0029.
35. Raisa E. D. Rudge, Elke Scholten, and Joshua A. Dijksman, "Advances and Challenges in Soft Tribology with Applications to Foods," *Sensory Science and Consumer Perception • Food Physics & Materials Science* 27 (2019): 90–97, https://doi.org/10.1016/j.cofs.2019.06.011.
36. C. Cornelio et al., "Mechanical Behaviour of Fluid-Lubricated Faults," *Nature Communications* 10, no. 1 (2019): 1274, https://doi.org/10.1038/s41467-019-09293-9; C. Cornelio and M. Violay, "Effect of Fluid Viscosity on Earthquake Nucleation," *Geophysical Research Letters* 47, no. 12 (2020), https://doi.org/10.1029/2020GL087854.
37. Gary B. Chapman and Giles R. Cokelet, "Flow Resistance and Drag Forces Due to Multiple Adherent Leukocytes in Postcapillary Vessels," *Biophysical Journal* 74, no. 6 (1998): 3292–3301, https://doi.org/10.1016/S0006-3495(98)78036-1.
38. "Atherosclerosis," Cleveland Clinic, accessed May 14, 2024, https://my.clevelandclinic.org/health/diseases/16753-atherosclerosis-arterial-disease.
39. Caroline Cheng et al., "Atherosclerotic Lesion Size and Vulnerability Are Determined by Patterns of Fluid Shear Stress," *Circulation* 113, no. 23 (2006): 2744–53, https://doi.org/10.1161/CIRCULATIONAHA.105.590018.
40. Magali Billen, "Introduction to Geophysics: 4.1 The Forces Driving Plate Motions," LibreTexts Geosciences, University of California, Davis, n.d., https://geo.libretexts.org/Courses/University_of_California_Davis/GEL_056%3A_Introduction_to_Geophysics/Geophysics_is_everywhere_in_geology . . . /04%3A_Plate_Tectonics/4.01%3A_The_Forces_Driving_Plate_Motions.
41. Junlin Hua et al., "Asthenospheric Low-Velocity Zone Consistent with Globally Prevalent Partial Melting," *Nature Geoscience* 16, no. 2 (2023): 175–181, https://doi.org/10.1038/s41561-022-01116-9.
42. Sunyoung Park et al., "Weak Upper-Mantle Base Revealed by Postseismic Deformation of a Deep Earthquake," *Nature* 615, no. 7952 (2023): 455–460, https://doi.org/10.1038/s41586-022-05689-8.
43. Jim McHugh, "Albert Kingsbury: His Life and Times," *Sound and Vibration,* October 2003, https://www.kingsbury.com/pdf/albert_kingsbury.pdf.
44. P. M. Cann, "The 'Leaves on the Line' Problem—a Study of Leaf Residue Film Formation and Lubricity under Laboratory Test Conditions," *Tribology Letters* 24 (2006): 151–158; Michael Watson et al., "The Composition and Friction-Reducing Properties of Leaf Layers," *Proceedings of the Royal Society A* 476, no. 2239 (2020): 20200057.

45. Joseph Jaffe et al., "'A Novel Methodology for Developing Ultra-Low Adhesion Leaf Layers on a Full-Scale Wheel / Rail Rig,'" *Proceedings of the Institution of Mechanical Engineers, Part F: Journal of Rail and Rapid Transit* 238, no. 6 (2024): 736–741, https://doi.org/10.1177/09544097231216521.

46. D. W. Wang et al., "How Do Grooves on Friction Interface Affect Tribological and Vibration and Squeal Noise Performance," *Tribology International* 109 (2017): 192–205, https://doi.org/10.1016/j.triboint.2016.12.043; Chris Baraniuk, "Tiny Bubbles Under a Ship May Be the Secret to Reducing Fuel Consumption," *Smithsonian*, November 13, 2020, https://www.smithsonianmag.com/innovation/tiny-bubbles-under-ship-may-be-secret-to-reducing-fuel-consumption-180976278/; Chang-Jin Kim, "Smooth Sailing," Physics World, May 18, 2017, https://physicsworld.com/a/smooth-sailing/.

47. Phong A. Tran and Thomas J. Webster, "Understanding the Wetting Properties of Nanostructured Selenium Coatings: The Role of Nanostructured Surface Roughness and Air-Pocket Formation," *International Journal of Nanomedicine* 8 (2013): 2001–9, https://doi.org/10.2147/IJN.S42970.

48. Kim, "Smooth Sailing."

49. Kim, "Smooth Sailing"; Ivan U. Vakarelski et al., "When Superhydrophobicity Can Be a Drag: Ventilated Cavitation and Splashing Effects in Hydrofoil and Speed-Boat Models Tests," *Colloids and Surfaces A: Physicochemical and Engineering Aspects* 628 (2021): 127344, https://doi.org/10.1016/j.colsurfa.2021.127344.

50. Richard Stimson, "The Power to Fly," n.d.; "Guide to Reliving the Wright Way," Beginners Guide to Aeronautics, Glenn Research Center, NASA, n.d., https://www1.grc.nasa.gov/beginners-guide-to-aeronautics/re-living-the-wright-way/.

51. Otto Lilienthal, *Birdflight as the Basis of Aviation: A Contribution towards a System of Aviation, Compiled from the Results of Numerous Experiments Made by O. and G. Lilienthal,* trans. from 2nd ed. by W. Isenthal (London: Longman's Green, 1911).

52. Lilienthal, *Birdflight as the Basis of Aviation;* Bjorn Fehrm, "Bjorn's Corner: Aircraft Drag Reduction, Part 2," Leeham News and Analysis, October 27, 2017, https://leehamnews.com/2017/10/27/bjorns-corner-aircraft-drag-reduction-part-2/.

53. Lilienthal, *Birdflight as the Basis of Aviation.*

54. "Wind Tunnel," The Franklin Institute, March 8, 2014, https://fi.edu/en/science-and-education/collection/wind-tunnel.

55. Fred E. C. Culick and Henry R. Jex, "Aerodynamics, Stability and Control of the 1903 Wright Flyer," Report WF 84/09-1, AIAA Wright Flyer Project, September 20, 1984, https://authors.library.caltech.edu/records/vea6k-35x13; Orville Wright and Wilbur Wright, Flying-machine, United States US821393A, filed March 23, 1903, and issued May 22, 1906, https://patents.google.com/patent/US821393A/en.

56. "Winglets Evolve to Boost Efficiency Across Aircraft Spectrum," *Aviation Week,* August 8, 2024, https://aviationweek.com/mro/aircraft-propulsion/winglets-evolve-boost-efficiency-across-aircraft-spectrum.

57. Sergey L. Chernyshev, Sergey V. Lyapunov, and Andrey V. Wolkov, "Modern Problems of Aircraft Aerodynamics," *Advances in Aerodynamics* 1, no. 1 (2019): 1–15.

58. Loz Blain, "Shark-Skin-Inspired Film Immediately Drops Airliner Fuel Consumption," New Atlas, February 23, 2022, https://newatlas.com/aircraft/aeroshark-aircraft

-skin/; "Reducing Skin-Friction Drag by Laminar Flow," *Aerospace Engineering Blog* (blog), April 28, 2012, https://aerospaceengineeringblog.com/skin-friction-drag/.

59. Steven Seman, "Lesson 6: Surface Patterns of Pressure and Wind," METEO 3: Introductory Meteorology, Pennsylvania State University, n.d.,https://www.e-education.psu.edu/meteo3/l6.html.

60. Bob Henson, "Hurricane Winds at Landfall: Why Is It They Seem to Fall Short?" Weather Underground, April 12, 2018, https://www.wunderground.com/cat6/hurricane-winds-landfall-why-it-they-seem-fall-short.

61. Russell E. Johnson, "Estimation of Friction of Surface Winds in the August 1949, Florida Hurricane," *Monthly Weather Review* 82, no. 3 (March1954): 73–79, https://doi.org/10.1175/1520-0493(1954)082<0073:EOFOSW>2.0.CO;2; Vance A. Myers, "Surface Friction in a Hurricane," *Monthly Weather Review* 87, no. 8 (August 1959): 307–311, https://doi.org/10.1175/1520-0493(1959)087<0307:SFIAH>2.0.CO;2.

62. "NOAA Saildrone Hurricane Observation," Pacific Marine Environmental Lab, NOAA, n.d., https://www.pmel.noaa.gov/saildrone-hurricane/.

63. Anders Persson, "User Guide to ECMWF Forecast Products," European Centre for Medium-Range Weather Forecasts, 2011, https://ghrc.nsstc.nasa.gov/uso/ds_docs/tcsp/tcspecmwf/ECMWFUserGuideofForecastProductsm32.pdf.

64. Renzhi Jing et al., "Global Population Profile of Tropical Cyclone Exposure from 2002 to 2019," *Nature* 626, no. 7999 (2024): 549–554, https://doi.org/10.1038/s41586-023-06963-z.

65. John P. Rafferty, "Hurricane Ian," Britannica, https://www.britannica.com/event/Hurricane-Ian-2022.

66. John Atkinson et al., "Deriving Frictional Parameters and Performing Historical Validation for an ADCIRC Storm Surge Model of the Florida Gulf Coast," *Florida Watershed Journal* 4 (2011).

67. Kyra M. Bryant and Muhammad Akbar, "An Exploration of Wind Stress Calculation Techniques in Hurricane Storm Surge Modeling," *Journal of Marine Science and Engineering* 4, no. 3 (2016), https://doi.org/10.3390/jmse4030058; Muhammad K. Akbar, Simbarashe Kanjanda, and Abram Musinguzi, "Effect of Bottom Friction, Wind Drag Coefficient, and Meteorological Forcing in Hindcast of Hurricane Rita Storm Surge Using SWAN + ADCIRC Model," *Journal of Marine Science and Engineering* 5, no. 3 (2017), https://doi.org/10.3390/jmse5030038.

5 · *A Waste of Energy*

1. Energy Institute, "Statistical Review of World Energy," 2024, https://www.energyinst.org/statistical-review/home.

2. Frequently Asked Questions, U.S. Energy Information Administration (EIA)," https://www.eia.gov/tools/faqs/faq.php?id=1394&t=6.

3. Robert W. Carpick et al., "Tribology Opportunities for Enhancing America's Energy Efficiency," report to the Advanced Research Projects Agency-Energy at the U.S. Department of Energy, February 2017, https://www.stle.org/images/PDF/STLE_ORG/whitepaper/Opportunities_for_Enhancing_Energy.pdf.

4. Kenneth Holmberg, Peter Andersson, and Ali Erdemir, "Global Energy Consumption Due to Friction in Passenger Cars," *Tribology International* 47 (2012): 221–234, https://doi.org/10.1016/j.triboint.2011.11.022.

5. Mathew Mate and Robert W. Carpick, *Tribology on the Small Scale* (Oxford: Oxford Graduate Texts, 2019).

6. Holmberg, Andersson, and Erdemir, "Global Energy Consumption."

7. N. Morris et al., "Combined Numerical and Experimental Investigation of the Micro-Hydrodynamics of Chevron-Based Textured Patterns Influencing Conjunctional Friction of Sliding Contacts," *Proceedings of the Institution of Mechanical Engineers, Part J: Journal of Engineering Tribology* 229, no. 4 (2015): 316–335, https://doi.org/10.1177/1350650114559996.

8. Kenneth Holmberg and Ali Erdemir, "The Impact of Tribology on Energy Use and CO2 Emission Globally and in Combustion Engine and Electric Cars," *Tribology International* 135 (2019): 389–396, https://doi.org/10.1016/j.triboint.2019.03.024.

9. L. Gaines, E. Rask, and G. Keller, "Which Is Greener: Idle, or Stop and Restart?" Argonne National Laboratory, US Department of Energy, 2012, https://afdc.energy.gov/files/u/publication/which_is_greener.pdf.

10. Jesse Crosse, "Stop-Start Systems: Is There a Long-Term Impact on My Car's Engine?" Autocar, April 5, 2022, https://www.autocar.co.uk/car-news/new-cars/stop-start-long-term-impact-your-car-s-engine.

11. "Nanotechnology Repairs Engine Damage in Cars," *NASA Tech Briefs*, August 1, 2020, https://spinoff.nasa.gov/Spinoff2020/cg_4.html; Crosse, "Stop-Start Systems."

12. Letícia Raquel de Oliveira et al., "Scuffing Resistance of Polyalphaolefin (PAO)-Based Nanolubricants with Oleic Acid (OA) and Iron Oxide Nanoparticles," *Materials Today Communications* 31 (2022): 103837, https://doi.org/10.1016/j.mtcomm.2022.103837.

13. Uzair Ahmad et al., "Biolubricant Production from Castor Oil Using Iron Oxide Nanoparticles as an Additive: Experimental, Modelling and Tribological Assessment," *Fuel* 324 (2022): 124565, https://doi.org/10.1016/j.fuel.2022.124565.

14. Alison Downing et al., "Castor Oil Plant," *Plant of the Week* (blog), School of Natural Sciences, Macquarie University, n.d., https://www.mq.edu.au/__data/assets/pdf_file/0004/1215346/Plant-of-the-week-Ricinus-communis-Castor-Oil-Plant.pdf.

15. Holmberg and Erdemir, "The Impact of Tribology."

16. Holmberg and Erdemir, "The Impact of Tribology."

17. Andrea Aiken, "Evolution of Copper Corrosion Testing for Electric Vehicle Lubricants," *TLT Magazine*, January 2024, https://www.stle.org/files/TLTArchives/2024/01_January/Feature.aspx.

18. Ashutosh Mishra, "Impact of Silica Mining on Environment," *Journal of Geography and Regional Planning* 8, no. 6 (2015): 150–156, https://doi.org/10.5897/JGRP2015.0495.

19. Mayura Lolage et al., "Green Silica: Industrially Scalable & Sustainable Approach towards Achieving Improved 'Nano Filler—Elastomer' Interaction and Reinforcement in Tire Tread Compounds," *Sustainable Materials and Technologies* 26 (2020): e00232, https://doi.org/10.1016/j.susmat.2020.e00232.

20. Lolage et al., "Green Silica"; Carpick et al., "Tribology Opportunities."

21. D. W. Wang et al., "How Do Grooves on Friction Interface Affect Tribological and Vibration and Squeal Noise Performance," *Tribology International* 109 (2017): 192–205, https://doi.org/10.1016/j.triboint.2016.12.043; J. Le Rouzic et al., "Friction-Induced Vibration by Stribeck's Law: Application to Wiper Blade Squeal Noise," *Tribology Letters* 49, no. 3 (2013): 563–572, https://doi.org/10.1007/s11249-012-0100-z.

22. Holmberg and Erdemir, "The Impact of Tribology."

23. Carpick et al., "Tribology Opportunities."

24. Myriam Harnafi et al., "A Simplified Model for the Wear Prediction of Plain Bearings in the Variable Stator Vane System," *Tribology International* 196 (2024): 109667, https://doi.org/10.1016/j.triboint.2024.109667.

25. Addison Schonland, "The Evolution of the Pratt & Whitney Geared Turbofan Engine," AirInsightGroup, March 22, 2016, https://airinsight.com/evolution-pratt-whitney -geared-turbofan-engine/.

26. Betty McCoy, "The Future of Jet Engine Lubrication," *TLT Magazine*, October 2021, https://www.stle.org/files/TLTArchives/2021/10_October/Feature.aspx.

27. Carpick et al., "Tribology Opportunities."

28. Kelsey Reichmann, "To Lower Emissions, the Military Focuses on Increasing Aircraft Efficiency," Avionics International, August 11, 2021, https://www.aviationtoday.com /2021/08/11/lower-emissions-military-focuses-increasing-aircraft-efficiency/.

29. Ali Altar Inceer, "Low Bypass Ratio Variable Cycle Engine Concepts for High Speed Aircraft" (MA thesis,, Chalmers University of Technology, Gothenberg, Sweden, 2022), https://odr.chalmers.se/items/af29630b-9bc9-4109-b52c-bae14b1c6ac8.

30. Holmberg and Erdemir, "The Impact of Tribology."

31. Okada Norio et al., "The 2011 Eastern Japan Great Earthquake Disaster: Overview and Comments," *International Journal of Disaster Risk Science* 2, no. 1 (2011): 34–42, https://doi.org/10.1007/s13753-011-0004-9.

32. Kenneth Holmberg and Ali Erdemir, "Influence of Tribology on Global Energy Consumption, Costs and Emissions," *Friction* 5, no. 3 (2017): 263–284, https://doi.org/10 .1007/s40544-017-0183-5.

33. Ian M. Hutchings and Philip Shipway, *Tribology: Friction and Wear of Engineering Materials,* 2nd ed. (Elsevier Science, 2017).

34. Holmberg and Erdemir, "Influence of Tribology on Global Energy."

35. Ivan G Rice, "Combined Cycles: Now at 62%, next 65%," *Turbomachinery Magazine,* March 11, 2017, https://www.turbomachinerymag.com/view/the-evolution-of-the -combined-cycle-power-plant-ii.

36. ASTM, "Standard Practice for Calculating Viscosity Index from Kinematic Viscosity at 40°C and 100°C," https://www.astm.org/d2270-10r16.html.

37. Clarion Energy Content Directors, "CCGT Power Plants on Track to Add 7.8 GW of Capacity This Year," *Power Engineering* (blog), November 4, 2022, https://www.power -eng.com/gas/ccgt-power-plants-on-track-to-add-7-8-gw-of-capacity-this-year/.

38. Clarion Energy Content Directors, "CCGT Power Plants."

39. Adina Popa, Rhodri Edwards, and Indran Aandi, "Carbon Capture Considerations for Combined Cycle Gas Turbine," *Energy Procedia,* 10th International Conference on

Greenhouse Gas Control Technologies, 4 (January 1, 2011): 2315–23, https://doi.org/10.1016/j.egypro.2011.02.122.

40. K. V. Sterkhov et al., "A Zero Carbon Emission CCGT Power Plant and an Existing Steam Power Station Modernization Scheme," Energy 237 (2021): 121570, https://doi.org/10.1016/j.energy.2021.121570.

41. IEA, "Natural Gas Is Now Stronger than Ever in the United States Power Sector—Analysis," International Energy Agency, December 4, 2023, https://www.iea.org/commentaries/natural-gas-is-now-stronger-than-ever-in-the-united-states-power-sector.

42. Mark Lammey, "Future Looks Bright for Global Subsea Engineering," Energy Voice (blog), February 5, 2019, https://www.energyvoice.com/events/subsea-expo/191951/future-looks-bright-for-global-subsea-engineering/.

43. Lawrence W. de Leeuw et al., "Modulating Pipe-Soil Interface Friction to Influence HPHT Offshore Pipeline Buckling," Ocean Engineering 266 (2022): 112713, https://doi.org/10.1016/j.oceaneng.2022.112713.

44. Carpick et al., "Tribology Opportunities."

45. Carpick et al., "Tribology Opportunities."

6 · Frontiers of Friction

1. Martin Ekman, "A Concise History of the Theories of Tides, Precession-Nutation and Polar Motions (from Antiquity to 1950)," Surveys in Geophysics 14 (1993): 585–617.

2. George Howard Darwin, "Tides," in Encyclopedia Britannica (1902 Encyclopedia), https://www.1902encyclopedia.com/T/TID/tides.html; William M. Kaula, "Tidal Dissipation by Solid Friction and the Resulting Orbital Evolution," Reviews of Geophysics 2, no. 4 (1964): 661–685, https://doi.org/10.1029/RG002i004p00661; Gordon J. F. MacDonald, "Tidal Friction," Reviews of Geophysics 2, no. 3 (1964): 467–541, https://doi.org/10.1029/RG002i003p00467.

3. George Howard Darwin, "Marriages Between First Cousins in England and Their Effects," Journal of the Statistical Society of London 38, no. 2 (1875): 153–184, https://doi.org/10.2307/2338660.

4. William B. Ashworth, "George Howard Darwin," Scientist of the Day (blog), July 9, 2018, https://www.lindahall.org/about/news/scientist-of-the-day/george-howard-darwin; George Howard Darwin, "1879 XIII. On the Precession of a Viscous Spheroid, and on the Remote History of the Earth," Philosophical Transactions of the Royal Society of London 170 (December 1879): 447–538, https://doi.org/10.1098/rstl.1879.0073.

5. George Howard Darwin, "1880 I. On the Secular Changes in the Elements of the Orbit of a Satellite Revolving about a Tidally Distorted Planet," Proceedings of the Royal Society of London 30, no. 200–205 (December 1880): 1–10, https://doi.org/10.1098/rspl.1879.0076; MacDonald, "Tidal Friction"; Sylvio Ferraz-Mello, Adrián Rodríguez, and Hauke Hussmann, "Tidal Friction in Close-in Satellites and Exoplanets: The Darwin Theory Re-Visited," Celestial Mechanics and Dynamical Astronomy 101, no. 1 (2008): 171–201, https://doi.org/10.1007/s10569-008-9133-x.

6. Walter Munk, "Once Again-Tidal Friction," *Quarterly Journal of the Royal Astronomical Society* 9 (1968): 352.

7. H. Gerstenkorn et al., "The Importance of Tidal Friction for the Early History of the Moon," *Proceedings of the Royal Society of London. Series A. Mathematical and Physical Sciences* 296, no. 1446 (1997): 293–297, https://doi.org/10.1098/rspa.1967.0023; Stanton J. Peale, P. Cassen, and Ray T. Reynolds, "Melting of Io by Tidal Dissipation," *Science* 203, no. 4383 (1979): 892–894.

8. Amirhossein Bagheri et al., "The Tidal–Thermal Evolution of the Pluto–Charon System," *Icarus* 376 (2022): 114871; Rory Barnes, "Tidal Locking of Habitable Exoplanets," *Celestial Mechanics and Dynamical Astronomy* 129 (2017): 509–536.

9. Gary D. Egbert and Richard D. Ray, "Significant Dissipation of Tidal Energy in the Deep Ocean Inferred from Satellite Altimeter Data," *Nature* 405, no. 6788 (2000): 775–778.

10. Kevin Zahnle et al., "Emergence of a Habitable Planet," *Space Science Reviews* 129 (2007): 35–78.

11. Zahnle et al., "Emergence of a Habitable Planet."

12. René Heller et al., "Habitability of the Early Earth: Liquid Water under a Faint Young Sun Facilitated by Strong Tidal Heating Due to a Closer Moon," *PalZ* (2021): 1–13; Jure Japelj, "How Much Did the Moon Heat Young Earth?" *Eos,* January 11, 2022, http://eos .org/articles/how-much-did-the-moon-heat-young-earth.

13. A. McEwen, K. de Kleer, and R. Park, "Does Io Have a Magma Ocean?" *Eos,* October 18, 2019, http://eos.org/features/does-io-have-a-magma-ocean; Japelj, "How Much Did the Moon Heat Young Earth?"; Peale, Cassen, and Reynolds, "Melting of Io by Tidal Dissipation."

14. Peale, Cassen, and Reynolds, "Melting of Io by Tidal Dissipation"; McEwen, de Kleer, and Park, "Does Io Have a Magma Ocean?"; Japelj, "How Much Did the Moon Heat Young Earth?"

15. Emeline Bolmont et al., "Solid Tidal Friction in Multi-Layer Planets: Application to Earth, Venus, a Super Earth and the TRAPPIST-1 Planets-Potential Approximation of a Multi-Layer Planet as a Homogeneous Body," *Astronomy & Astrophysics* 644 (2020): A165.

16. Héctor Aceves and Maria Colosimo, "Dynamical Friction in Stellar Systems: An Introduction," *arXiv Preprint Physics / 0603066,* 2006; R. J. Tayler, "Subrahmanyan Chandrasekhar. 19 October 1910–21 August 1995," *Biographical Memoirs of Fellows of the Royal Society* 42 (1996): 81–94.

17. Tayler, "Subrahmanyan Chandrasekhar."

18. Carl Sagan, *The Demon-Haunted World* (New York: Ballantine Books, 1997).

19. Freeman Dyson, "Chandrasekhar's Role in 20th-Century Science," *Physics Today* 63, no. 12 (2010): 44–48.

20. T. Padmanabhan, "Stellar Dynamics and Chandra," *Current Science* 70, no. 9 (1996): 784–788.

21. Tayler, "Subrahmanyan Chandrasekhar"; "The Nobel Prize in Physics 1983," NobelPrize.org, https://www.nobelprize.org/prizes/physics/1983/chandrasekhar/facts/.

22. Jaco de Swart, "Deciphering Dark Matter: The Remarkable Life of Fritz Zwicky," *Nature* 573, no. 7772 (2019): 32–34.

23. Lachlan Lancaster et al., "Dynamical Friction in a Fuzzy Dark Matter Universe," *Journal of Cosmology and Astroparticle Physics* 2020, no. 01 (2020): 1; Héctor Velázquez and Simon D. M. White, "Sinking Satellites and the Heating of Galaxy Discs," *Monthly Notices of the Royal Astronomical Society* 304, no. 2 (1999): 254–270.

24. A. Trelles et al., "Concurrent Infall of Satellites-Collective Effects Changing the Overall Picture," *Astronomy & Astrophysics* 668 (2022): A20; Simon D. M. White, "Simulations of Sinking Satellites," *Astrophysical Journal*, 274 (1983): 53–61; Velázquez and White, "Sinking Satellites and the Heating of Galaxy Discs."

25. Man Ho Chan and Chak Man Lee, "Indirect Evidence for Dark Matter Density Spikes around Stellar-Mass Black Holes," *Astrophysical Journal Letters* 943, no. 2 (2023): L11; Robert Lea, "Black Holes May Be Swallowing Invisible Matter That Slows the Movement of Stars," Space.com, March 27, 2023, https://www.space.com/black-holes-may-be-swallowing-invisible-matter.

26. Lea, "Black Holes May Be Swallowing Invisible Matter"; Chan and Lee, "Indirect Evidence for Dark Matter Density Spikes."

27. Y. B. Zel'dovich and I. D. Novikov, "The Hypothesis of Cores Retarded during Expansion and the Hot Cosmological Model," *Soviet Astronomy* 10 (1967): 602–603.

28. "Did Black Holes Form Immediately after the Big Bang?" European Space Agency, December 16, 2021, https://www.esa.int/Science_Exploration/Space_Science/Did_black_holes_form_immediately_after_the_Big_Bang.

29. Jane H. MacGibbon and B. J. Carr, "Cosmic Rays from Primordial Black Holes," *Astrophysical Journal* 371 (1991): 447, https://doi.org/10.1086/169909.

30. Bernard Carr and Florian Kühnel, "Primordial Black Holes as Dark Matter Candidates," *SciPost Physics Lecture Notes*, 2022, 048.

31. Tim Stephens and Mike Peña, "A Wobble from Mars Could Be Sign of Dark Matter, New Study Finds," UC Santa Cruz News, September 17, 2024, https://news.ucsc.edu/2024/09/primordial-black-holes.html; Tung X. Tran et al., "Close Encounters of the Primordial Kind: A New Observable for Primordial Black Holes as Dark Matter," *Physical Review D* 110, no. 6 (September 17, 2024): 063533, https://doi.org/10.1103/PhysRevD.110.063533.

32. Adam Mann, "Are Tiny Black Holes Hiding within Giant Stars?" *Science*, December 13, 2023, https://www.science.org/content/article/are-tiny-black-holes-hiding-within-giant-stars.

33. Matthew E. Caplan, Earl P. Bellinger, and Andrew D. Santarelli, "Is There a Black Hole in the Center of the Sun?" *Astrophysics and Space Science* 369, no. 1 (2024): 8, https://doi.org/10.1007/s10509-024-04270-1; Monisha Ravisetti, "Is a Black Hole Stuck inside the Sun? No, but Here's Why Scientists Are Asking," Space.com, January 5, 2024, https://www.space.com/black-hole-stuck-inside-sun; John G. Cramer, "A Black Hole in Our Sun?" Alternate View column, *Analog Science Fiction and Fact Magazine*, May–June 2024, https://www.npl.washington.edu/av/altvw230.html.

34. Earl P. Bellinger et al., "Solar Evolution Models with a Central Black Hole," *Astrophysical Journal* 959, no. 2 (2023): 113, https://doi.org/10.3847/1538-4357/ad04de.

35. Masamune Oguri, Volodymyr Takhistov, and Kazunori Kohri, "Revealing Dark Matter Dress of Primordial Black Holes by Cosmological Lensing," *Physics Letters B* 847 (2023): 138276, https://doi.org/10.1016/j.physletb.2023.138276.

36. Marc Oncins et al., "Primordial Black Holes Capture by Stars and Induced Collapse to Low-Mass Stellar Black Holes," *Monthly Notices of the Royal Astronomical Society* 517, no. 1 (2022): 28–37.

37. Ákos Szölgyén, Morgan MacLeod, and Abraham Loeb, "Eccentricity Evolution in Gaseous Dynamical Friction," *Monthly Notices of the Royal Astronomical Society* 513, no. 4 (2022): 5465–73.

38. Pravin Kumar Natwariya, Jitesh R. Bhatt, and Arun Kumar Pandey, "Viscosity in Cosmic Fluids," *European Physical Journal C* 80, no. 8 (2020): 767.

39. Daniel Reiche, Francesco Intravaia, and Kurt Busch, "Wading through the Void: Exploring Quantum Friction and Nonequilibrium Fluctuations," *APL Photonics* 7, no. 3 (2022); Mário G. Silveirinha, "Theory of Quantum Friction," *New Journal of Physics* 16, no. 6 (2014): 063011; Fiona MacDonald, "Physicists Say They've Manipulated 'Pure Nothingness' and Observed the Fallout," ScienceAlert, January 19, 2017, https://www .sciencealert.com/physicists-say-they-ve-managed-to-manipulate-pure-nothingness.

40. Nikita Kavokine, Marie-Laure Bocquet, and Lydéric Bocquet, "Fluctuation-Induced Quantum Friction in Nanoscale Water Flows," *Nature* 602, no. 7895 (February 2022): 84–90, https://doi.org/10.1038/s41586-021-04284-7; Eleonora Secchi et al., "Massive Radius-Dependent Flow Slippage in Carbon Nanotubes," *Nature* 537, no. 7619 (2016): 210–213, https://doi.org/10.1038/nature19315; Tim Wogan, "New Phenomenon 'Quantum Friction' Explains Water's Bizarre Properties," *Chemistry World,* February 2022, https://www.chemistryworld.com/news/new-phenomenon-quantum-friction-explains -waters-bizarre-properties/4015163.article.

41. Chen Ly, "Quantum Friction Explains Strange Way Water Flows through Nanotubes," *New Scientist,* February 2, 2022, https://www.newscientist.com/article/2306900-quan tum-friction-explains-strange-way-water-flows-through-nanotubes/.

42. John B. Pendry, "Quantum Friction–Fact or Fiction?" *New Journal of Physics* 12, no. 3 (2010): 033028; Thomas G. Philbin and Ulf Leonhardt, "No Quantum Friction between Uniformly Moving Plates," *New Journal of Physics* 11, no. 3 (2009): 033035.

43. Reiche, Intravaia, and Busch, "Wading through the Void."

44. Pendry, "Quantum Friction–Fact or Fiction?"; Reiche, Intravaia, and Busch, "Wading through the Void."

45. Alex Evilevitch, "The Mobility of Packaged Phage Genome Controls Ejection Dynamics," ed. William Gelbart and Arup K. Chakraborty, *eLife* 7 (2018): e37345, https://doi .org/10.7554/eLife.37345.

46. Evilevitch, "Mobility of Packaged Phage Genome"; Andrea Soranno et al., "Integrated View of Internal Friction in Unfolded Proteins from Single-Molecule FRET, Contact Quenching, Theory, and Simulations," *Proceedings of the National Academy of Sciences* 114, no. 10 (2017): E1833–39, https://doi.org/10.1073/pnas.1616672114.

47. I. S. Gabashvili and A. Y. Grosberg, "Dynamics of Double Stranded DNA Reptation from Bacteriophage," *Journal of Biomolecular Structure & Dynamics* 9, no. 5 (1992): 911–920, https://doi.org/10.1080/07391102.1992.10507966.

48. Ting Liu et al., "Solid-to-Fluid–like DNA Transition in Viruses Facilitates Infection," *Proceedings of the National Academy of Sciences* 111, no. 41 (2014): 14675–80, https://doi .org/10.1073/pnas.1321637111.

49. Evilevitch, "The Mobility of Packaged Phage Genome."

50. Mounir Fizari et al., "Role of DNA-DNA Sliding Friction and Nonequilibrium Dynamics in Viral Genome Ejection and Packaging," *Nucleic Acids Research* 51, no. 15 (2023): 8060–69, https://doi.org/10.1093/nar/gkad582.

51. Davide Marenduzzo et al., "Topological Friction Strongly Affects Viral DNA Ejection," *Proceedings of the National Academy of Sciences* 110, no. 50 (2013): 20081–86, https://doi.org/10.1073/pnas.1306601110.

52. Fizari et al., "Role of DNA-DNA Sliding Friction."

53. Soranno et al., "Integrated View of Internal Friction."

54. Soranno et al., "Integrated View of Internal Friction"; Troy Cellmer et al., "Measuring Internal Friction of an Ultrafast-Folding Protein," *Proceedings of the National Academy of Sciences* 105, no. 47 (2008): 18320–25, https://doi.org/10.1073/pnas.0806154105.

55. Sandhyaa Subramanian et al., "Slow Folding of a Helical Protein: Large Barriers, Strong Internal Friction, or a Shallow, Bumpy Landscape?" *Journal of Physical Chemistry B* 124, no. 41 (2020): 8973–83, https://doi.org/10.1021/acs.jpcb.0c05976; David de Sancho, Anshul Sirur, and Robert B. Best, "Molecular Origins of Internal Friction Effects on Protein-Folding Rates," *Nature Communications* 5, no. 1 (2014): 4307, https://doi.org/10.1038/ncomms5307; Soranno et al., "Integrated View of Internal Friction"; Stephen J. Hagen, "Solvent Viscosity and Friction in Protein Folding Dynamics," *Current Protein & Peptide Science* 11, no. 5 (2010): 385–395, https://doi.org/10.2174/138920310791330596.

56. Soranno et al., "Integrated View of Internal Friction."

57. Soranno et al., "Integrated View of Internal Friction."

58. Soranno et al., "Integrated View of Internal Friction."

Acknowledgments

AS I HOPE TO have conveyed, tribology today stands on the shoulders of giants like Newton himself, and I am just one tribologist who has hoped to capture the magic of the tribology community. To all the tribologists I've had the great fortune of meeting throughout my career: this is for you and because of you. I hope I've done you justice.

While it is impossible to name all the tribologists who have impacted my life, please humor me as I shout out to those in my immediate tribology family. The tribology lab I trained in at the University of Florida has recently moved on, as Greg Sawyer has executed a marvel of a pivot from tribology to cancer research. But my fellow tribology grad students from those days have become lifelong professional partners. A particular thanks to Rachel Thayer and her weekly sanity-check phone calls. I escaped the heat of Florida to the (seriously) delightful weather of the UK and the extraordinary kindness and leadership of Rob Dwyer-Joyce at the Leonardo Center for Tribology at the University of Sheffield. I still heed his uncle's advice and wear silly socks on important occasions. And finally, I thank my Imperial College London family, who still greet me like an old friend when I show up at the college unannounced: Philippa Cann, an absolute legend who will be annoyed I've said that, Daniele Dini, and Marc Masen, who managed to find me a physical copy of the Jost Report (hero!), thank you for welcoming me to the fold. And to Connor, Louise, and Suzy—I'm spoiled to have such an amazing tribology family.

I would not even have considered this feat without the people who urged me to write a book and have probably regretted it every day since: the "Jcrew" of Jonathan Plant and Jonathan Hempfling. Because J names seem to flock together, I also couldn't have done this without

Dr. Jessica Allen, the best kind of friend I could ask for, stemming all the way back from our first university physics course. To Maria, thank you for always bringing the fire. Rob and Lizzi, I appreciate your visits more than you realize. And of course, a huge shout-out to the Delaware Institute of Tribology. That WD-40 mug helped power this book.

My family has had to tolerate my random tribological facts and writing moods these past months while I tried to figure out how to tell this story. Thank goodness for Tom Marks, for keeping me and everything around us going. What a gift to this world he is. To my mom, Paula, who somehow knew this was possible even when I didn't. To my dad, David, and to Barbara: I'm ready for a long vacation out west, please! I'm looking forward to resuming silly and great exchanges with my brother, Jason, over science fiction movies and books. Thank you, Abby, for putting up with us! I hope this book makes my niece and nephew proud; Lilia and Davis, you keep inspiring me to reach for my dreams, so thank you! And, obviously, I have to acknowledge the TriboCats, who helped me through various stages of this trek: Snickers, Mocha, and Latte. My scrapbook of the writing process is much cuter because of those purrballs.

For bringing complex science to life in the illustrations throughout this book, thank you to David R. Nicholas. Your artistic skills did what mine could not!

And, finally, I thank the woman who made this happen, my brilliant editor Rachel Field, who saw the potential for this book just from a TED Talk. Thank you for helping me get this book across the finish line.

Thank you to all my readers. Stay curious, always.

Index

Page numbers in *italic* indicate illustrations.